High-Pressure Pumps

High-Pressure Pumps

First Edition

Includes experience and research to help engineers, scientists and end users understand the technical side of pumps and systems for high-pressure applications.

Michael T. Gracey, P.E.
Weatherford International
Houston, TX

AMSTERDAM • BOSTON • HEIDELBERG • LONDON • NEW YORK • OXFORD
PARIS • SAN DIEGO • SAN FRANCISCO • SINGAPORE • SYDNEY • TOKYO

Gulf Professional Publishing is an imprint of Elsevier

G P
P

Gulf Professional Publishing is an imprint of Elsevier
30 Corporate Drive, Suite 400, Burlington, MA 01803, USA
Linacre House, Jordan Hill, Oxford OX2 8DP, UK

∞ Recognizing the importance of preserving what has been written, Elsevier prints its books on
acid-free paper whenever possible.

Library of Congress Cataloging-in-Publication Data
Application Submitted

British Library Cataloguing-in-Publication Data
A catalogue record for this book is available from the British Library.

ISBN 13: 978-0-7506-7900-8
ISBN 10: 0-7506-7900-X

For information on all Gulf Professional Publishing
publications visit our Web site at www.books.elsevier.com

06 07 08 09 10 11   10 9 8 7 6 5 4 3 2 1

Printed in the United States of America

Working together to grow
libraries in developing countries

www.elsevier.com | www.bookaid.org | www.sabre.org

ELSEVIER   BOOK AID   Sabre Foundation
                International

Contents

Chapter 8 – Stripping and Surface Preparation _____ 113

Chapter 9 – Environmental and Safety Concerns and Improvements _____ 127

Chapter 10 – Hot Water Washdown Unit _____ 155

Chapter 11 – Troubleshooting High-Pressure Pumps _____ 167

Chapter 12 – High-Pressure Pump Systems _____ 197

Preface

This book is to document the development of high-pressure pumps and systems technology that is being used in almost every industrial endeavor. High-pressure pumps and ultra-high pressure equipment is being used for hydrostatic testing, erosion testing, surface preparations, cutting material, machining metals and cleaning surfaces. The new technology is used in the aerospace industry, petrochemical-related fields, steel mills, automotive plants and food processing. It has been found effective to reduce the cost and improve results over other methods of accomplishing a given task. The experiences and research discussed herein will help engineers, scientists, and end users to understand the technical side of pumps, nozzles, accessories and power pump systems that have been developed for special applications. With this background, more uses can be found for the technology so that the pump industry can continue to grow in the future.

Michael T. Gracey

Acknowledgments

The high-pressure pump industry as a whole and each of us individually would not be where we are today if not for the ones who came before us. My very job depends on a man named C.J "Cobe" Coberly, who started working with pumps over 90 years ago. Men like R.G. LeTourneau, who said, "I'm just a mechanic that God used," inspired me to enter engineering by his example of what can be achieved by one individual. Original ideas are rare—we learn from those around us and teach what we have learned to others. Chapter 1 is the acknowledgment of people who helped make the high-pressure pump industry what it is today. They left a legacy of innovative design, good engineering practices, and integrity to inspire us. Further acknowledgment is given to those people who have contributed to the subject at the end of each chapter. Thanks to Oscar Hernandez and especially to Emil Levek for many of the CAD drawings used in the book.

Michael T. Gracey

For by Him were all things created, that are in heaven, and that are in earth, visible and invisible, whether they be thrones, or dominions, or principalities, or powers: all things were created by Him, and for Him.
Colossians 1:16

About the Author

Mike Gracey was born in Port Arthur, TX, in 1942, and grew up around the oil refineries & chemical plants in that area. His early fascination with mechanisms like engines and transmissions led him to pursue an engineering education in order to understand "how things work." After studying at LeTourneau College in Longview, TX, he attended night school at Lamar University in Beaumont, TX, while working as an apprentice pipefitter in a shipyard, where he eventually moved into the engineering department. After returning to full-time studies, he worked as a licensed tankerman in the Port Arthur and Beaumont areas until graduating in 1971. By 1973, he was working at the National Maritime Research Center (NMRC) in Galveston, TX, and became involved in testing specialized equipment and methods for water-blast cleaning, surface preparation, and coatings. This led him to the more specialized area of high-pressure pumps used for water jetting. Starting in 1976, he designed and built systems for high-pressure pump manufacturers and packagers in Texas, Maryland, and Michigan. Currently, he works for Weatherford International, Inc., using Kobe's line of industrial pumps to handle chemicals, water, oil products, hot fluids, and cryogenics. He has published over 40 articles and technical papers.

Chapter 1

History of High-Pressure Pumps

1.0   Introduction to High-Pressure Pumps

A reciprocating pump can be defined as a mechanical device that consists of one or more single- or double-acting positive-displacement elements (pistons or plungers) that imparts a pulsating dynamic flow to a liquid. This definition also explains that the pistons or plungers are driven in a more or less harmonic motion by a rotating crank with a connecting rod arrangement. This motion generates flow by pulling the fluid through inlet check valves and pushes the fluid through outlet check valves that are located near the inlet and outlet of the pump.

An alternative definition describes pumps as devices for exerting pressure on fluids for transportation or {through them to transmit pressure to a more or less remote point where it is transformed into work.} For positive-displacement, high-pressure, and piston or plunger pumps, it could be said that they create flow (not pressure) until the flow is restricted, which, in turn, causes the pressure to increase in the fluid.

1.1   Early History of Pumps

Since the beginning of civilization, there has been a need to move water from one place to another. Cupped hands have given way to clay vessels, wooden buckets, and aqueducts to provide water to remote locations.

The earliest mechanical device of authentic record for lifting water was called the *shadoff*. The Egyptians used this device as early as 1500 BC for watering their herds and irrigating farmland, which is depicted in Fig. 1.1.

The shadoff consists of a counterweighted, pivoted pole with a rope and bucket that the operator uses to his advantage to draw water from a water source, such as the Nile River. This type of device was used on *The Amazing Race* (CBS, August 14, 2004), when the contestants had to draw water from the Nile and fill a one liter jar, then ride a donkey across the fields to a village where they filled a two liter clay pot. It took the contestants at least two trips to accomplish the task. The Chinese may have made the next improvement to this system by attaching buckets at intervals to a loop of rope over a windlass that could be turned by hand or treadmill. This device, the *Chinese continuous bucket rope*, is depicted in Fig. 1.2 and enabled the villagers to move water from a river into the rice fields for irrigation.

Around the beginning of the Christian Era, the Romans extended the rotary principle by creating the *Roman bucket wheel*. Buckets were attached to a large wheel that would spin and dip each bucket into the water. The wheel then lifted the buckets and dumped the water into aqueducts, as illustrated in Fig. 1.3.

1.2   The Force Pump

A Greek engineer named Ctesibius is believed to be the inventor of the *force pump*. This pump had two vertical cylinders mounted side by side with single-acting pistons. A walking beam actuated these pistons so the pumps provided a practically continuous stream. Today, Weatherford International manufactures the Ram Pump to handle multiphase oil products, and it is powered by hydraulic cylinders, as shown in Fig. 1.4.

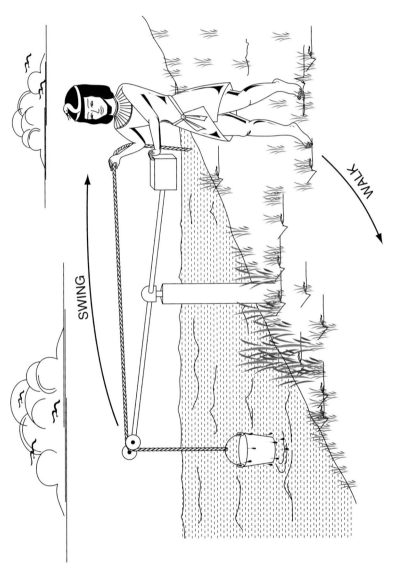

Figure 1.1. Shadoff used to lift water by Egyptians circa 1500 BC

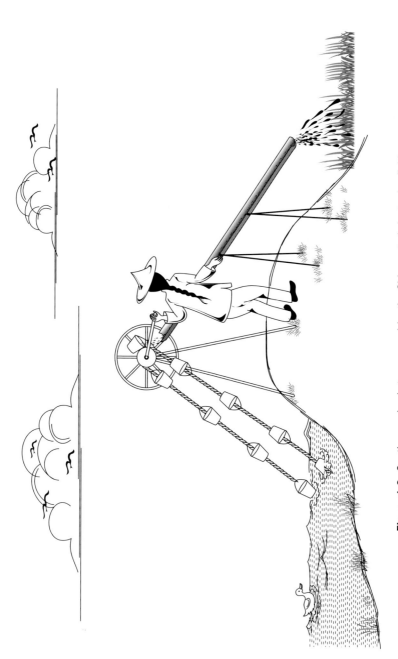

Figure 1.2. Continuous bucket rope used by the Chinese to irrigate rice fields

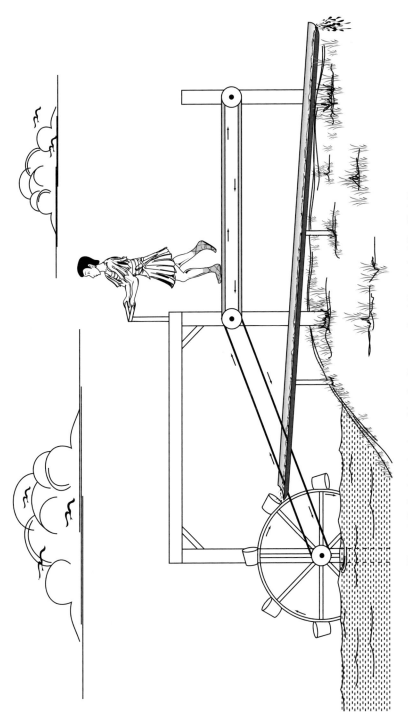

Figure 1.3. Bucket wheel used by the Romans about the start of the Christian era

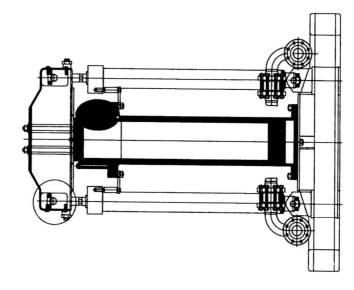

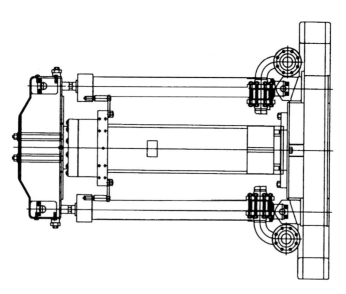

Figure 1.4. Ram Pump developed for multiphase pumping

1.3 The Screw Pump

The screw principle of raising water is generally credited to another Greek inventor, Archimedes. The origin of modern progressive cavity-pumping equipment, such as the *screw pump*, can be traced to his ancient designs. Giambattista Della Porta is credited with suggesting the use of steam acting directly on the surface of the water in 1601. The cost and physical limitations of dewatering mines with animal-powered pumps inspired the use of steam, but it was too wasteful. The steam condensed before it had done its work, so in 1707 French scientist Denis Papin proposed separating the steam and water by a piston. This system reduced the condensation and provided a pocket in the piston that could be preheated by inserting a hot iron block before the steam was applied. In America, Henry R. Worthington invented the first direct-acting, reciprocating steam pump in 1840, which widened the field of steam pumping applications. Prior to this invention, there were no small steam pumps, so the new design was employed in Erie Canal steamboats until that type of boat was no longer used. The pumps were then removed from those boats and used for another 30 years for other purposes.

1.4 The Centrifugal Pump

The first known use of centrifugal force to pump liquid was designed by Johann Jordan in 1680, and the first pump resembling the modern centrifugal pump was designed in 1818 by an unknown inventor. This latter pump is called the *Boston* or *Massachusetts pump*, a name that suggests its birthplace. The rotary pump design also traces its history back to the gear-and-lobed type found in a collection made by a Frenchman named Serviere who was born in 1593. The sliding van-type rotary pump was described by an Italian named Agostino Ramelli in a book that was published in 1588. More recently, in 1915, Adolph Wahle of Davenport, IA, patented the turbine or regenerative pump. Steam pumps may still be in use today, but they have become secondary to power-driven pumps because of the efficiency of the direct-acting pump and the various means of driving it. Power can be provided by electric motor or engine and be connected by a belt drive, chain drive, or gear box. The pump unit can be portable when engine-driven or be powered by electricity over great distances for a wide field of applications. The history of high-pressure pumps continues in several countries and often involves interesting needs and individuals.

1.5   United States Pumps

Sir Samuel Moreland invented the plunger-type pump in 1675. While the piston (with a leather seal) in a cylinder had been used before the Christian Era, Moreland's pump may have been the first use of a piston rod and stuffing box (packing in a cylinder) to displace water. The history of high-pressure, positive-displacement pumps in the United States is closely tied to the oil industry. Whether it is a power-driven, double-acting, reciprocating pump used in the oil fields or a high-pressure water-jetting pump used for petrochemical cleaning, cutting, and surface preparation, this type of pump comes from and is used by the oil industry. The early models were used for crude oil gathering, artificial lift, pipeline applications, drilling mud, refining products, and salt-water disposal. The materials have improved; the valves and the packing designs, which come in vertical and horizontal models, have also evolved; the modern-day pump design continues to develop to provide efficiency and dependability for pumping fluids at high pressure.

1.5.1   Kobe Pumps

The Kobe firm was founded in 1923 by C. J. Coberly, born December 31, 1892, in Cameron, MO, the youngest of a family of pioneer cattlemen. Coberly graduated from Stanford University in 1915 with a degree in mechanical engineering. His associates called him "Cobe" and said that he never wavered in his faith of the merits of hydraulic pumping and never compromised his adherence to engineering principles, manufacturing quality, and ethical conduct. The *Kobe Pump,* with its vertical configuration in Size 2, Size 3, and Size 4 models–examples of an excellent design with precision quality—has survived even into the twenty-first century. Other U.S. pump companies include *Gardner-Denver, Union, Gaso,* and *Kacy.*

1.5.2   Union Pumps

The *David Brown Union Pump Company* had its inception in 1885 at the company's main production facility in Battle Creek, MI.

1.5.3   FMC Pumps

Food Machinery Corporation (FMC) Technologies and its pumps
have an interesting history that began when John Bean moved to
California from Hudson, MI, in 1883. *The Growing Orbit*, published
by FMC in 1992, is a colorful record of the FMC story. Bean's insecti-
cide pump was patented in 1884 and later led to the creation of the
Bean Pump Company. Bean's grandson, David C. Crummey, became
the first president of the Bean Pump Company in 1888 and is credited
with developing the company that, after a series of mergers, is now
known by its initials, FMC.

1.5.4   Kerr Pumps

Kerr Pump is based in Sulfur, OK, and has been manufacturing high-
pressure piston and plunger pumps since 1946. The company's product
line includes simplex, duplex, triplex, quintuplex, and sextuplex models.

1.6   German Pumps

The *Hammelmann Pump* is a vertical positive-displacement pump
that, along with the *Kamat, Woma*, and *Wepuko* pumps, originated in
Germany.

1.6.1   Wepuko Pumps

Wepuko Hydraulik has been producing high-pressure pumps and
compressors since 1932, although the roots of the company can be traced
back to the mid-19th century. In 1864, Heinrich Esche established a
clothing manufacturing company on Wepuko's current site, and the
very first building from that time is still standing. In 1888, a mechanic
named Cluck started a machine shop in nearby Bad Urach, which was
purchased in 1910 by Wilhelm von Neudeck. In 1912, Von Neudeck, with
a partner named Haas, purchased the pump business of Gustav
Magenwirth. Even after the purchase of the Metzingen clothing factory
in 1917, Von Neudeck and Haas continued to manufacture pumps and
tools both in Metzingen and Bad Urach until the partners' acrimonious
breakup in 1922. Haas kept the machine shop in Bad Urach and
continued to make pumps. The company later became known as Uraca.
Von Neudeck continued pump manufacturing in Metzingen until selling

the company to Fritz Thumm in 1932. Thumm had been Von Neudeck's shop foreman and was trained at Uraca. He began manufacturing pumps and compressors in Metzingen and renamed the company Wepuko Hydraulik in 1955 (an acronym for *we*rkzeug [tools] *pu*mpen [pumps], and *ko*mpressor [compressors]). An early hand pump is shown in Fig. 1.5, and a triplex pump is shown in Fig. 1.6 that was made around 1935.

1.6.2 Uraca Pumpenfabrik GmbH & Co. KG

Uraca was founded in Bad Urach in 1893. As of this writing, Uraca is a privately owned company that employs approximately 260 people with an export rate of about 60% of production. In addition to sales offices in Germany, Uraca has a subsidiary in Paris and offices in Thailand and Dubai. The main activity of the company includes research, design, and production of high-pressure plunger pumps up to 40,000 psi (2,800 bar) and power ratings to 2,000 hp (1,500 kW). The

Wepuko Hand Pump
1935

Figure 1.5. Hand-operated pump circa 1935 (*Courtesy of Wepuko*)

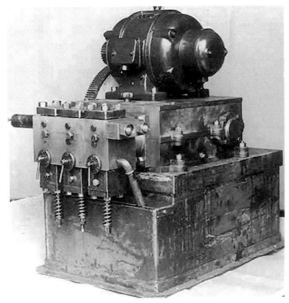

Wepuko Triplex Pump
1935

Figure 1.6. Early triplex pump circa 1935 (*Courtesy of Wepuko*)

pumps are used for press drives, descaling, boiler feed, and also to supply oil to hydrostatic bearing systems. Uraca Pumps are also part of high-pressure cleaning systems.

1.7   Japanese Pumps

1.7.1   Sugino Corporation

Sugino Machine was founded in 1936 to produce air-driven tube cleaners under the name Sugino Cleaner Works in Osaka, Japan. Because of heavy bombing during World War II, the factory moved to Uozu, Toyama, in 1945. Rinpei Sugino was the first president of Sugino Cleaner Works, Ltd, which was established in 1956. In 1967, the company launched an automatic self-feeding pneumatic drill line and established a branch office in Chicago, IL, in 1975. Around 1980, the water-blast industry became aware of 2-D and 3-D rotating nozzles that Sugino had developed. The company also advertised high-pressure pumps and deburring equipment, but the export pricing may have been a problem in the American market. By 1994, Sugino developed the Jet-Flex machining and deburring system, and in 1993, U-Jet cavitation technology was introduced to the United States. Sugino builds pumps that are in the 200 Mega pascals (MPa) pressure range to the ultra-high-pressure (UHP) range of 394 MPa.

1.7.2   Other Japanese Pumps

Other pumps may be produced in Japan, but only the Cat Pump series and pressure washer-size positive-displacement pumps are known in the U.S. market as of this writing.

1.8   Chinese Pumps

1.8.1   Wuxi Petrel Pump Company

A company in China named Wuxi Petrel started manufacturing pumps in 1992 and now produces 18 series of pumps for various applications. The company's high-pressure pumps have improved in quality, and Wuxi acquired International Organization for Standardization (ISO)-9001 certification for its quality management system in 1999. By 2002, the company had finished the construction of a new Haiyan manufacturing plant and maintained its ISO certification. One of the high-pressure pumps produced by Wuxi is shown in Fig. 1.7.

1.8.2   YongTai Pump Company

The YongTai Pump Company started production of high-pressure pumps and metering pumps around 1984. The company first acquired ISO-9002 certification in May 2001. Since 2000, a new facility has been completed for managing the overflow work; another plant was completed in October 2004. An example of the company's high-pressure pumps is shown in Fig. 1.8.

1.9   Later History of Pumps

Old-line pumps were pressed into service for a larger variety of products that needed to be moved at increasingly higher pressures. As early as the 1930s, the Aldrich Pump Company (acquired by Ingersoll-Rand in 1961) was combining high-pressure pumps with nozzles for steel mill descaling. An early user of oil-field pumps was Fred Machol,

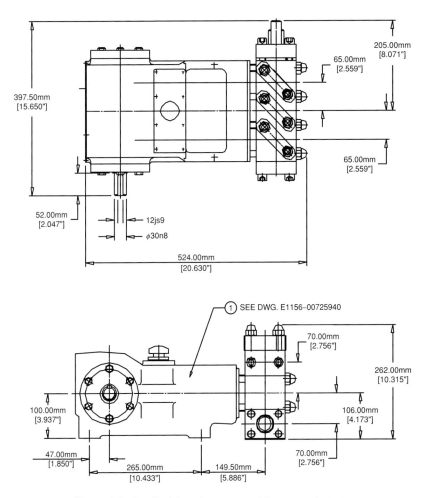

Figure 1.7. Small triplex piston pump (*Courtesy of Wuxi*)

who later started Acme Cleaning Equipment in 1966. Machol was a contract cleaner in 1949 when he was asked to clean the wooden slats of a cooling tower for Standard Oil Refinery in Cleveland, OH. He rented a pump that produced 300 psi at 20 gpm to successfully remove the oily dirt from the cooling tower slats. In 1951, Solvent Services (co-founded by Machol) was asked by Republic Steel to clean the checkers of the open-hearth furnaces (checkers are brick latticework over which hot air from the furnaces is passed and then reversed to recapture the heat when needed). Machol purchased a John Bean pump with a 20-horse-power, two-cylinder gasoline engine to supply water to a homemade

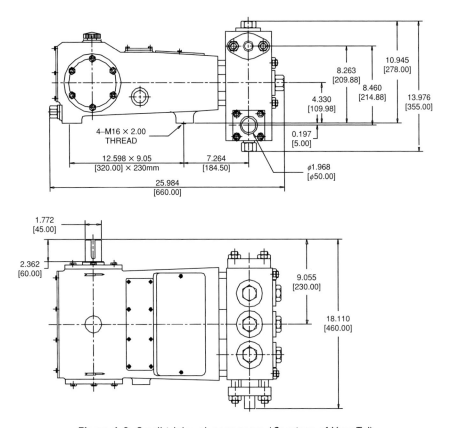

Figure 1.8. Small triplex plunger pump (*Courtesy of YongTai*)

lance that took five men to handle. The lance nozzle was a capped piece of pipe with holes drilled around it to make an early tube nozzle. The system cleaned the dirt and soot from the brick so well that the lettering was readable, and the superintendent was very pleased with the new method. Because the cleaning process took only 12 hours instead of the usual 20 hours, Republic gave Machol's company enough work to justify the building of two more pump units. Machol made them from Wilson Snyder pumps and Ford industrial engines mounted on trucks. Nine or ten cleaning lances were operated from one pump without a control valve or foot-gun, so every operator had to control his lance to avoid hurting someone.

An engineer and salesman named John Goss reported that by the mid-1950s, the available pressure for positive displacement pumps was as high as 3,000 to 5,000 psi, but these pumps were mainly used and built by independent cleaning contractors. Goss explained that there was talk of using high-pressure water for cleaning, but it was hard to find a pump that would last more than about 2 hours before it had to be repaired. Dick Paseman, who was working as an industrial paint contractor, reported that one of the biggest breakthroughs came in 1956 when an engineer from Monsanto in St. Louis, MO, came to Houston, TX, to speak at the local chapter of The Society of Painting Contractors (SPC). He promoted the use of high-pressure water to clean and desalt steel before painting because of Houston's proximity to the Gulf of Mexico.

The real boost for high-pressure industrial cleaning originated with a coating problem that occurred in the local refineries. Mastic was commonly used in the late 1940s and early 1950s to coat tanks and valves to prevent having to paint the steel. By the late 1950s, the mastic product began to fail when pinholes allowed the steel substrate to rust and corrode. The thick mastic coating had to be removed, but scraping was a slow, labor-intensive process and sandblasting did not work successfully. Lou Sline, whose Sline Industrial Painters was one of the largest industrial paint contractor companies in the South, experimented with an idea presented at the coating society; in 1956, Sline used high-pressure water to remove the mastic coating from refinery equipment along the Gulf Coast. From Sline's experience, Paseman found other uses for high-pressure water jetting when a foreman at one of the refineries asked him if a water blaster could clean heat exchangers. Paseman tried using a Graco airless paint spray hose with a nozzle to lance the heat exchanger tubes, and this method worked.

In 1959, Goss and Paseman formed the American Powerstage Company, which developed and copyrighted a pump unit called a Water Blaster by 1961. The first design used a 10-piston wobble plate pump driven by an engine that could produce over 7,000 psi, but the first year of its use was reported to be a nightmare because of the difficulty keeping the equipment running. American Powerstage switched to a Wheatley pump, which worked better, but Goss and Paseman then changed brands when another pump company approached them. American Powerstage finally standardized on a Union pump when pump reliability was still only about 3 hours. One of the most successful water-blasting pump companies in the mid-1960s was Chemical Cleaning, Inc. (CCI); this company advanced high-pressure pump technology but was limited by the accessories available to meet the pump's capabilities. Cracked fluid blocks were still a problem for CCI and other early equipment manufacturers.

1.9.1   Job Master

Job Master, a company started by Clem Mundy in the 1960s, built high-pressure water blasters with Gardner-Denver Pumps (PE-5) driven by diesel engines (6v53 Detroit) with transmissions, chain drive, and air-operated clutches. These units were mounted on custom trailers with fuel tanks built into the frames. Figure 1.9 shows an early trailer-mounted water blaster using a Gardner-Denver PE-5 pump. In lower gears, the pump could handle up to 10,000 psi, but this pressure was probably not guaranteed by Gardner-Denver Pumps.

1.9.2   Partek

Maurice Wolf came to Houston, TX, around 1964. An avid golfer, Wolf founded Partek (from the golfing term "par", and "tek" for technical). Wolf's brother-in-law, Lou Sline, gave him the idea to form Partek to build water-blasting equipment. The early water-blast

Figure 1.9. Early portable pump unit for water blasting circa 1965

units used John Bean pumps mounted on trailers and were powered by diesel engines. By 1968, Partek had developed a stainless steel fluid-end design, but a fluid block cracking problem developed, and Wolf seemed reluctant to invest more money. Partek's example, however, inspired other entrepreneurs. An employee named Roy Groneaur left Partek around 1965 and used a Wheatley pump and fluid-end versions similar to the Wheatley block-style started Tritan Corporation. George Rankin left the company and started Aqua-Dyne using in-line valving for his pump that eliminated the large fluid block. Partek recovered and developed a more dependable 10,000-psi pump that helped the company's success in the 1970s and 1980s. After its acquisition by Butterworth, with Mike Ginn as president, the former Partek group started marketing the company's 40,000-psi pump in the 1990s under the name Butterworth Jetting Systems, as shown in Fig. 1.10.

1.9.3   Aqua-Dyne

From the 1970s to the time of this writing, Aqua-Dyne, under Rankin's leadership, continued to develop the in-line valve arrangement and the accessories needed to service the developing markets for water-blasting equipment.

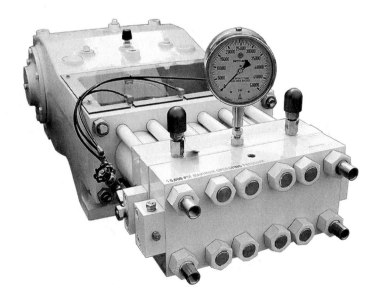

Figure 1.10. Triplex plunger pump for 40,000-psi operation *(Courtesy of Butterworth)*

1.9.4   Hydro-Services

Companies like Hydro-Services began using water-blasting equipment for maintenance cleaning around 1962 in Missouri City, TX, under the leadership of Jack Hinrichs. Equipment was purchased from Gardner-Denver and Tritan Corporations, which would often use Wheatley pumps. In 1979, Hydro-Services negotiated an Original Equipment Manufacturer (OEM) contract with the Wheatley Pump Company in Tulsa, OK, and pumps were purchased from Wheatley by the new manufacturing division. The pumps (150 and 75) had block-style fluid ends with tapered seated valves to operate up to 6,000 psi. Until this time, it was difficult to find pipefittings, hoses, and accessories that could handle 10,000 psi or more of pressure. When J. D. Frye moved from Tritan to Hydro-Services, he developed a cartridge valve design that enabled Hydro-Services to produce its own fluid ends to fit the Wheatley and Myers power frames. Figure 1.11 shows the Dura-Pak valve cartridge. After Hinrichs' retirement in 1978, the company was

Figure 1.11. Valve cartridge design developed by Hydro-Services

led by Jim Terry as president, Pat Debusk and Joe Parish as vice-presidents, and John Hinrichs as in-house legal counsel. The company was sold to Tracor in 1985 and became HydroChem in the 1990s.

1.9.5   National Liquid Blaster

Forest Shook drove his pickup truck to Battle Creek, MI, in 1964 to buy his first pump to be used in a water-blast unit, which would mark the start of his new business. This unit was used in a Ford Motor Company plant to clean paint from production equipment. From his garage near Wixom, MI, Shook founded National Liquid Blaster (NLB), which now has offices in four cities in the United States.

1.9.6   Jetech

Jerry Desantis purchased a machine shop in Michigan in the 1980s to start building large pumps and systems. He developed an ultra-high-pressure (UHP) fluid-end that was combined with several power frames, including Myers, Union, and Wilson-Snyder, to deliver up to 40,000 psi for the company's water-jetting service. The pump and accessories developed by Jetech are marketed worldwide.

1.9.7   Flow Industries

Flow Industries has stated that Norman C. Franz, a pioneer in ultra-high-pressure water jetting and a former professor of forestry at the University of British Columbia (Franz patented a concept for a very high-pressure water-jet cutting system in 1968), has long been regarded as the father of the water jet. As a forestry engineer who wanted to find new ways to slice thick trees into lumber, Franz was the first person to study the use of UHP water as a cutting tool in the 1950s. To create a UHP effect, Franz dropped heavy weights onto columns of water, forcing that water through a tiny orifice. He obtained short bursts of very high pressures, in some cases pressures higher than what is currently being used. From these bursts of high pressures, Franz found that he could cut wood and other materials. His later studies involved more continuous streams of water, but he found it difficult to obtain continuous high pressures. In addition, component life of the cutting tool was measured in minutes—not weeks or months as is common in the twenty-first century.

Franz took his idea to McCartney Manufacturing, now owned by Ingersoll-Rand, and the company sold its first water-jet cutting system in 1971 to the Alton Box Board Company in Jackson, TN, to cut furniture shapes from 1/4-inch-thick laminated paper tubes that saws and routers could not handle very well. Though Franz never made a production lumber cutter, his research and findings were used by Flow International Corporation to develop and produce pure water jet machines powered by intensifier pumps in the early 1970s. Flow sold its first system for industrial manufacturing in 1974, and the first major application for this technology was cutting disposable diapers. In 1979, Flow began researching methods to increase the water-jet cutting power in order to cut metals and other hard materials. A team of researchers, led by Mohamed Hashish, invented the abrasive water jet. The team found that by entraining a fine mesh garnet abrasive (a mesh similar to the abrasive used on sandpaper) into the waterjet stream, the water jets could cut virtually any material. In the 1980s, water jets cut the first steel, with Flow obtaining the patent for the technology in 1983. Ironically, the wood-cutting application that Franz first researched in the early days is now just a minor application for UHP technology. Today, UHP water jet technology is used for a wide range of projects for a variety of industries. As technology evolves, new applications and uses for this technology are continually being developed.

1.9.8   Jet Edge

Chris Possis started Jet Edge in 1984 as part of the Possis Corporation to develop UHP water jet technology. Jet Edge is located in St. Michael, MN, and manufactures UHP water jet cutting and cleaning equipment for precision cutting, surface preparation, coating removal, and hydro-demolition. The company continues to expand its 55,000-psi (3,800 bar) at 1.1-gpm (4.1-L) UHP equipment to cut virtually any metallic or nonmetallic material.

1.9.9   New Jet Technologies

Michael G. Mullinaux worked with Ingersoll-Rand until around 1981 before joining Flow Research to work with 40,000-psi intensifiers for cutting sandstone and drilling holes in mining operations. Previously, most of the UHP equipment was manufactured for industrial production line operations and had never been used for mobile applications.

Figure 1.12. Ultra-high-pressure portable water jetting unit (*Courtesy of NewJet*)

Between 1983 and 1987, a new group called ADMAC was formed to develop portable systems and tools for abrasive jet cutting, heat exchanger tube cleaning, and surface preparation using rotary jettip lances. By 1988, ADMAC was resorbed into Flow Systems, which is now Flow International. As early as 1984, ADMAC personnel saw the need for large direct-acting pumps instead of intensifiers for pressures of 40,000 psi to be used for concrete scarifying. The intensifier fluid-end design was adapted to a conventional pump power frame, and Ingersoll-Rand features were added to produce the first NewJet UHP pumps in 1996 to 1997. Figure 1.12 shows the new triplex pump design for pressures up to 40,000 psi.

The companies described in this chapter and people involved with them have helped make the high-pressure pump industry what it is in the twenty-first century. They have left a legacy of innovative design, good engineering practices, and examples of integrity to inspire the high-pressure pump industry of the future.

References

Campbell, A., and Real, M. (1992). *Growing orbit: The story of FMC Corporation.* FMC Corporation, Chicago.

Coberly, C. J. (1961). *Theory and application of hydraulic oil well pumps.*

Chapter 2

Pump Design

2.0   Pump Definition

Positive displacement, reciprocating pumps can be classified as power pumps or direct-acting pumps; horizontal or vertical pumps; single-acting or double-acting pumps; piston, plunger, or diaphragm pumps; and simplex, duplex, or multiplex pumps. In the high-pressure and ultra-high-pressure (UHP) fields, intensifiers and plunger pumps are most often used to provide pressures in the range of 10,000–40,000 psi. A reciprocating pump is a positive displacement machine and not a kinetic machine like a centrifugal pump, so it does not require velocity to achieve pressure. It is an advantage to obtain high pressure at low velocity for large flows and slurry applications. A reciprocating pump has high efficiencies in the range of 85–94%, with a 10% loss through belts, gears, bearings, packing, and valves. *Flow capacity* is a function of pump speed and displacement and is relatively independent of pressure, but fluid can only be moved if the suction supply delivers fluid to the pump. In addition, a pump may not be able to suck liquid into itself because there is no tensile strength to the fluid, but it can remove the air from the pump cavity, which creates a partial vacuum to allow the pump chambers to fill.

2.1 Pump Net-Positive Suction Head

Net-positive suction head (NPSH) is the amount of head produced by the suction system and is expressed in meters of liquid, pounds per square inch, or feet. Net-positive suction head available (NPSHA) must exceed the net-positive suction head required (NPSHR) for proper operation of the pump and can be computed using the following equations:

$$\text{NPSHA} = Pa = hs - Pv - ha - hf \qquad \text{(Equation 1, for open or vented suction system)}$$

$$\text{NPSHA} = (hs - hf) - ha = hs - (hf + ha) \qquad \text{(Equation 2, for closed systems)}$$

In Equations 1 and 2, Pa represents atmospheric pressure, hs refers to static head, Pv equals vapor pressure, ha represents acceleration head, and hf refers to friction head. In a closed system, NPSHA can be taken as the difference between the static head and the sum of the friction losses and acceleration head. NPSHR is the amount of head in feet, pounds per square inch, meters, or bars absolute that is required at the suction inlet of the pump. NPSHR is related to losses in the suction manifold, pumping chamber, and suction valves. NPSHR does not include acceleration head, and the pump manufacturer determines the head by using tests outlined by the Hydraulic Institute.

The proper valve design greatly affects NPSH and should be selected on the basis of pressure, temperature, suction pressure, fluid compatibility, and pump speed. Pump speed and the ability of the valve to open and close properly (as affected by spring tension, valve weight, and valve area) have the greatest impact on NPSH. Different speeds and suction pressures affect how these factors act when the pump is operating.

2.2 Acceleration Head

The acceleration head is not constant because the liquid must accelerate and decelerate a number of times for each rotation of the crankshaft. The liquid inertia requires energy to produce acceleration, and the energy is returned to the system upon deceleration; adequate excess pressure must be provided to accelerate the liquid on the suction side of the pump to prevent cavitation. The crankshaft throws are located at 120-degree intervals in a triplex pump, and the plunger does not travel at a constant speed when the pump is running. The crossheads, connecting rods, and plungers accelerate and decelerate the liquid because the rotary motion of the crank is converted into linear motion with the maximum acceleration at 60, 180, and 300 degrees of rotation. To determine the effect of the action, the mass of the liquid in the suction line, and its acceleration, the force required is converted to pressure in feet of head. The Hydraulic Institute has provided the calculation in published standards as follows:

$$ha = \frac{LVnC}{gk} \qquad \text{(Equation 3)}$$

In Equation 3, *ha* represents acceleration head in feet, *L* is the actual length of the suction line in feet (not equivalent length), *V* is the velocity of liquid in the suction line, *n* refers to the rotating speed in revolutions per minute, *A* is area in square feet, *cfs* is cubic feet per second, and *C* is a constant that depends on pump type. The value for constant *C* depends on the type of pump. Unusual connection rod and crankshaft radius ratios can make these values change (.400 for simplex single-acting pump; .200 for simplex double-acting pump and duplex single-acting pump; .115 for duplex double-acting pump; .066 for triplex double-acting pump and duplex single-acting pump; .040 for quintuplex pump). The constant *K* refers to the compressibility of the liquid (1.4 for deaerated water; 2.5 for hydrocarbons with high compressibility, such as propane and methanol), and *g* is a gravitation constant of 32.2 ft/s^2.

The velocity for Equation 3 can be determined using the following formula:

$$V = \text{velocity of liquid in suction line, feet per second} \neq V \frac{cfs}{A}$$

2.3   Pump Crankshaft

The principle of crankshaft movement is fairly standard, with 10 degrees of crank movement at midstroke moving the piston or plunger 8% of its stroke. The 10 degrees of movement at the end of the stroke moves the piston or plunger 1% of its stroke. The rotation speed of the pump can have a drastic effect on the accelerated head because it increases as the square of the pump speed if the suction line length and velocity are held constant. A small high-speed pump does not have fewer pulsations than a large pump because all triplex pumps have a 25% variation in flow and pulse six times per revolution. Figure 2.1 is an example of a larger triplex power end for use up to 1,000 hp. On the figure, item 1 is a typical oil field skid to support the large fabricated power frame shown as item 4. Item 2 is the main crankshaft, and item 3 is the input shaft for the built-in gear reduction. Item 5 is the intermediate rod seal, item 6 is a splash shield, and item 7 is a coupler for the piston rod. Figure 2.2 shows a triplex power-end design (with a cast power frame and dual input shafts) that is typically used up to 500 hp.

The flow variations and pulses are as follows:

- Simplex pumps have 320% flow variation and 2 pulses per revolution.
- Duplex pumps have 160% flow variation and 4 pulses per revolution.
- Triplex pumps have 25% flow variation and 6 pulses per revolution.
- Quintuplex pumps have 8% flow variation and 10 pulses per revolution.

2.4   Pump Packing

Packing and plungers vary in type and material. For pressure ranges up to 4000 psi, the packing may be a square-braided type and is often

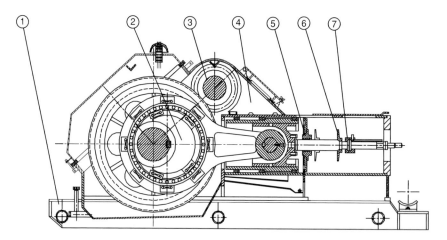

Figure 2.1. Power frame assembly for large pump

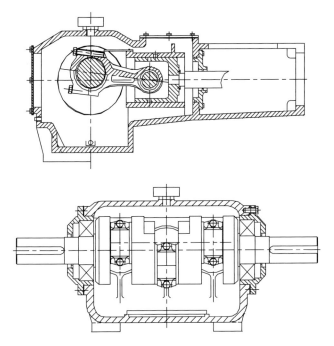

Figure 2.2. Typical triplex pump power frame

adjustable or nonspring loaded, as shown in Fig. 2.3. In this figure, item 1 is the plunger, item 2 is the packing adjustment nut, item 3 is the stuffing box, item 4 is a throat ring, item 5 is the square-braided packing, and item 6 is the stuffing box seal.

Rope packing is the term used for packing rings that are cut from a continuous spool. The rings can be made from cotton duct impregnated with carbon and from more modern materials, such as Teflon and Kevlar square-braided packing rings. Even though there are some exceptions, V-ring spring-loaded packing, as shown in Fig. 2.4, is often used for pressures over 5,000 psi to improve the sealing characteristics and eliminate the need to adjust the packing nut. In Fig. 2.4, item 1 is the plunger, item 2 is the packing retainer nut, and item 3 is the stuffing box. Item 5 is the packing spring that provides the preloading of item 6, and item 7 is the V-ring packing.

The stuffing box assembly in Fig. 2.4 has special features, such as item 8, which is a seal in the lantern ring, shown as item 9. Item 10 is a seal in the packing retainer nut to keep packing leakage from migrating along

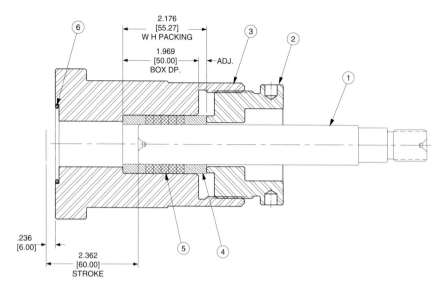

Figure 2.3. Square-braided packing used for lower pressure applications

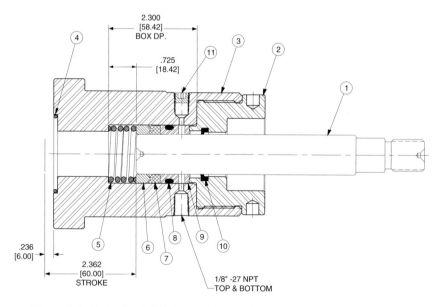

Figure 2.4. Spring-loaded V-ring packing used in medium pressure applications

the plunger. This box is used with packing leak detection instrumentation for pollution prevention, as described in Chapter 9.

The packing nuts for high-pressure pumps support the packing system when using adjustable packing and nonadjustable packing; however, the nut can be locked into a position by several methods. Locking methods include setscrews, clamps, cams, and short round bars inserted into holes in the packing nut. Plungers for positive displacement pumps are usually connected to the intermediate rod of the power end by threads, clamps, or collets. Some connections have been designed to self-align the plunger, which requires a button end, as shown in Fig. 2.5.

The plunger to intermediate rod connections includes threads on the plunger, flanges on both the intermediate rod and plunger, and several types of clamp designs. The collet design, which appeared around 1980, as shown in Fig. 2.6, is probably the newest innovation in the high-pressure pump field and requires only a plain end on the plunger for connection to the intermediate rod. The connector allows the use of coated plungers or plungers made from solid tungsten carbide and ceramic without threads. The idea of a collet design connection may have come from the machine shop method of holding tools with a tapered collet mechanism, but, in any case, the idea has been effectively adapted to high-pressure pump designs.

2.5 Pump Valves

Pump valves come in a variety of configurations, including those with tapered valve seats, O-ring sealed valve seats, in-line valves, hemispherical valves, ball valves, and valve cartridges. Figure 2.7 shows a standard

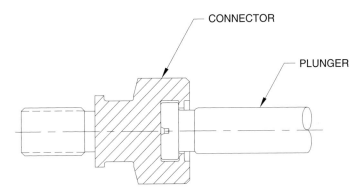

Figure 2.5. Slip-in plunger to intermediate rod connector

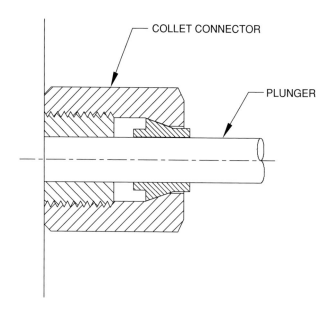

Figure 2.6. Collet-plunger to intermediate rod connector

tapered valve seat configuration with an insert on the valve body to improve sealing and efficiency. Item 1 is the tapered valve seat to match the taper in the pump fluid block. Item 2 is the wing-guided valve body with an elastomer insert (Item 3). Item 4 is the valve spring, item 5 is the valve cage, and item 6 is a locking O-ring between the seat and valve cage.

A disadvantage of the tapered valve seat is the tendency for it to become too tight for easy removal. When the valve seat cannot be removed by conventional means, heating the block is often the next step, which leads to fluid cylinder failure. Using mechanical pullers is the most common method for valve seat removal, but sometimes a hydraulic puller system is used successfully for hard-to-remove valves. The plain disk valves shown in Fig. 2.8 have their place in many pump applications. The disk can be made of materials such as Delrin, carbon steel, or titanium and used with tapered and O-ring sealed valve seats, which are easier to service and tend to be less expensive than for other valve designs.

Wing-guided poppet valves or flat-faced stem valves are more often used in the high-pressure pump designs. Figure 2.9 shows an in-line valve fluid-end design that successfully uses a wing-guided discharge valve and a disk suction valve. First developed for operation at

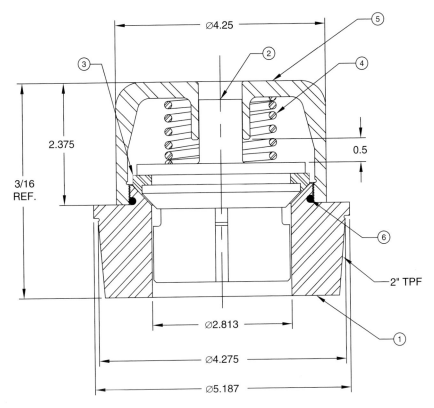

Figure 2.7. Valve assembly with tapered valve seat

20,000 psi, the fluid end runs well up to 30,000 psi, as experienced in field applications.

The pressure of 30,000 psi appears to have been ignored in the creation of high-pressure pumps because of the introduction of 36,000-psi intensifiers and the 40,000-psi pumps used for UHP water jetting. Few (if any) pump designers have concentrated on 30,000 psi, so the UHP designs can be used in this pressure range, when applicable and for better service life.

2.6 Fluid-End Design

Fluid-end design and components such as a fluid cylinders and stuffing boxes are designed with the working pressure and fluid to be pumped in mind. A fluid block with crossbores and tapered valve seats can be run at up to 10,000 psi at 30 gpm, for example, but the valve seats sometimes become almost impossible to extract, and

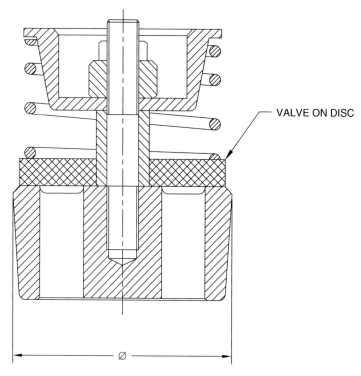

Figure 2.8. Valve assembly with disc-type valve

eventually the block will crack from fatigue. Several designs using O-ring-sealed valve seats and smaller fluid blocks can run from 10,000 to 20,000 psi without using the in-line valve configuration. Figure 2.10 shows a ball valve cartridge that can be used as a suction valve and a discharge valve. Item 1 is the valve housing, item 2 is the ball, item 3 is the valve spring, item 4 is the ball guide, and item 5 is a retainer ring for assembly and valve servicing.

This cartridge valve design can be held by the suction and discharge manifolds, as indicated by drawing of the fluid end shown in Fig. 2.11. The design allows the use of a single fluid block or individual blocks for a sectionalized or modularized fluid end.

The in-line valve fluid cylinder has proven to be the most effective design to eliminate the large crossbores and manifold problems. With in-line valves, a pressure range of 20,000–30,000 psi and above can be reached. Figure 2.12 shows an in-line valve arrangement that operates at pressures up to 40,000 psi. A list of items shown on the figure include:

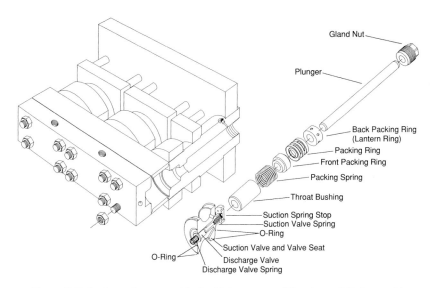

Figure 2.9. In-line valve assembly for high pressure (*Courtesy of Butterworth*)

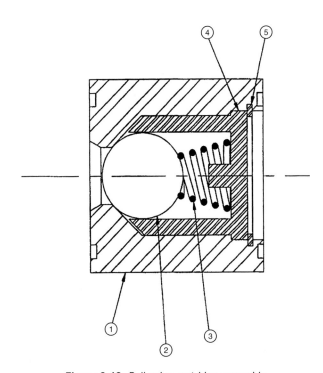

Figure 2.10. Ball valve cartridge assembly

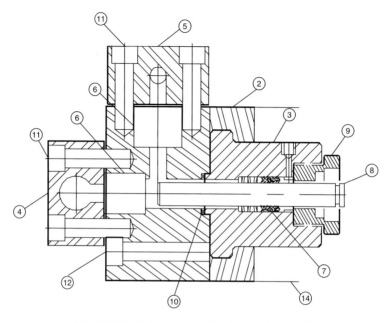

Figure 2.11. Ball valve cartridge fluid end assembly

1. Fluid block
2. Adapter plate
3. Stuffing box
4. Suction manifold
5. Discharge manifold
6. Valve assembly pocket
7. Packing set
8. Plungers
9. Packing nut
10. Stuffing box seal
11. Manifold cap screws
12. Fluid block cap screws
13. Adapter plate cap screws (not shown)
14. Pump power frame

The fluid-end design shown in Fig. 2.12 is rated with up to 40,000-psi working pressure. Its cylindrical fluid barrels, in-line valve arrangement, and UHP packing have overcome many of the problems associated with old fluid-end designs. Refer to Chapter 3 for more details on high-pressure fluid-end design and how it has evolved over the years.

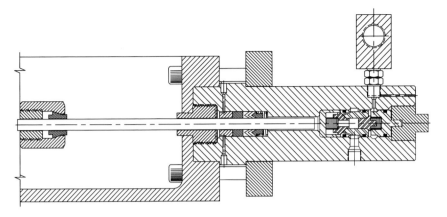

Figure 2.12. In-line valve assembly for ultra-high pressure (*Courtesy of Reliable Pumps, Inc.*)

Reference

Hydraulic Institute. *Hydraulic institute standards*. (1983) Hydraulic Institute: Cleveland, OH.

Chapter 3

Evolvement of High-Pressure Pumps

3.0   High-Pressure Pumps

High-pressure pumps evolved from the humble beginnings described in Chapter 1 because of the need to pump more types of fluid at ever-increasing pressures. Innovative people developed the fluid block/cylinder designs, the packing designs, and the valve designs to handle up to 10,000 psi, then 20,000 psi, then 36,000 psi, and then 40,000 psi. When each plateau was reached and the pump designs were barely debugged, someone was already taking further steps to reach the next plateau. Some of the modern ultra-high-pressure (UHP) pumps/intensifiers can even offer pressures up to and above 55,000 psi. Each breakthrough leads to the next plateau (where material, hoses, connections, and accessories have to catch up with the higher pump pressure). The design of high-pressure pumps has progressed to the standards of twenty-first century technology to handle the many fluids needed by modern industry.

3.1   Pump Power Ends

Power-end designs have been around for a long time and have been used by builders of high-pressure systems who could provide their own fluid ends to handle pressures of 10,000 psi, 15,000 psi, and 20,000 psi. Figure 3.1 shows a typical power frame for horizontal, positive displacement pumps. The material for building power frames for high-pressure pumps has improved over the years. For example, a forged crankshaft made from 1045 material broke at the journal radius. The rated rod load for the pump was 3969 lb (1,800 kg) for continuous service or 7,938 lb (3,600 kg) for intermittent service. The client attempted to run the pump at 2,000 psi and 80 gpm with 65 mm plungers, for a rod load of 10,286 lb (4,664 kg). The calculations revealed that the maximum principal stress of the crankshaft was being exceeded by 6,900 psi. The broken crankshaft material was tested and found to have a yield strength of 61,600 psi, which was too low for this application. The metallurgist recommended 4140 or 4340 if the crankshaft were locally hardened; if it were carburized, he suggested considering an 8620 or a 9310 material according to the design requirements. The 4140 material was chosen to replace the 1045 crankshaft material.

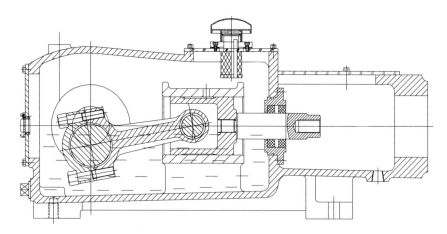

Figure 3.1. Horizontal positive displacement pump power end

In addition to greater horsepower pump units, high-pressure systems brought more revenue to contractors in the 1980 economic downturn when they offered 36,000 psi to clean hard-to-remove substances from heat exchangers and reactors. Material selection has played a part in improving pump power ends by using forged steel crankshafts instead of cast iron and by using better bearings for increased rod load capability.

Power frames and power ends of high-pressure pumps have slowly developed from the low-speed versions, such as the earlier *Gardner-Denver PE-5 Pumps,* to the high-speed versions, such as the Butterworth line of pumps. Some of the UHP manufacturers moved from the long stroke intensifier to fast turning triplex power ends for their mobile pump units. In municipal markets, for example, pumps are reaching 650 rpm, and some of the UHP pumps are reaching 2,000 rpm to produce the desired flow and pressure. While the power frame housing has traditionally been gray iron or nodular iron for pumps up to 500 hp, some larger pumps in the range of 600–1,000 hp are fabricated from mild steel. The earlier slow-turning pumps were driven with enclosed chain and sprockets, and the faster–turning versions were often belt driven to match the driver with the pump design speed. Separate gearboxes that had a two-speed feature were sometimes used, or an engine with a multispeed transmission was found useful for various applications. With a single-input shaft extension and double-input shaft extensions, some pumps have built-in gear reductions, while some pumps have built-on gear reductions, but sometimes a combination of gear reducer and belt drive reduction is needed to keep the high-pressure pump operating in the proper speed range. Figure 3.2 shows a power frame with a built-in gear; Item 1 is the power frame, and Item 2 is the fluid end. Figure 3.3 shows a built-on gear.

3.2　Pump Fluid Ends

Fluid-end design and components for some of the more modern pumps are the same as those of pumps built 50 years ago. The materials, however, may be selected to handle the fluid being pumped and the environment of the installation. A fluid end with tapered valve seats and large crossbores in the fluid block is limited to pressures less than 10,000 psi, as shown in Fig. 3.4. A fluid end with in-line valves and better materials is designed to handle up to 20,000-psi or more of pressure, as shown in Fig. 3.5.

In low-pressure pumps, the tapered valve seats are sometimes replaced with a straight bore valve seat that has an O-ring seal.

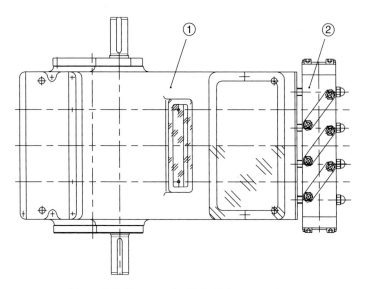

Figure 3.2. Power end with built-in gear reducer

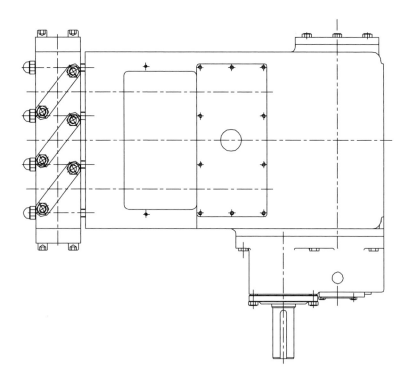

Figure 3.3. Power end with built-on gear reducer

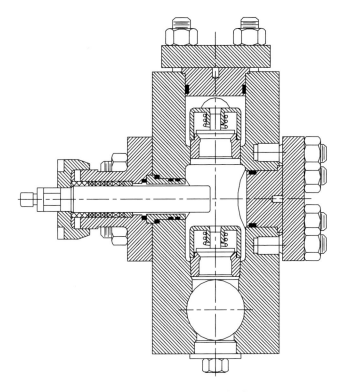

Figure 3.4. Fluid block with tapered valve seats

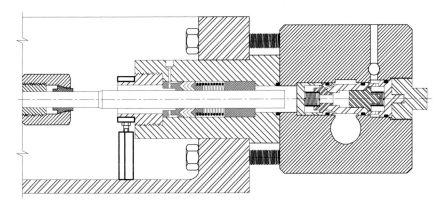

Figure 3.5. Fluid end with in-line valves developed by Reliable

Completing a calculation (as shown in Appendix D) for a fluid block can determine the pressure range for the fluid cylinder. According to the results of this calculation, the resulting pressure capability of the fluid block was in the range in which this small triplex pump typically operates.

An early design approach that eliminated the crossbores and the tapered valve seats used the individual fluid cylinders that contained the packing and in-line valves. Figure 3.6 shows a fluid-end design that was used by Kobe Pump Company in the 1940s to offer a pressure of 30,000 psi.

In the 1960s, packagers purchased John Bean, Union, and Wheatley pumps to build systems that could be operated from 6,000 to 10,000 psi and later designed their own fluid ends to solve some of the problems encountered with the existing designs. Figure 3.7 shows a block-style fluid end that a water-blasting contractor used until 1980 to reach pressures up to 10,000 psi.

By 1975, at least one of the high-pressure packaging companies was using an in-line valve fluid end that was adapted to Armco power frames and then to Gaso power frames, as shown in Fig. 3.8.

Fluid-end designs seemed to have developed because of the need for higher pressures in the water-blasting industry and in industrial

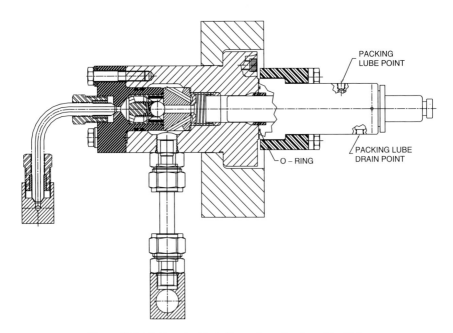

Figure 3.6. Fluid end with in-line valves developed by Kobe

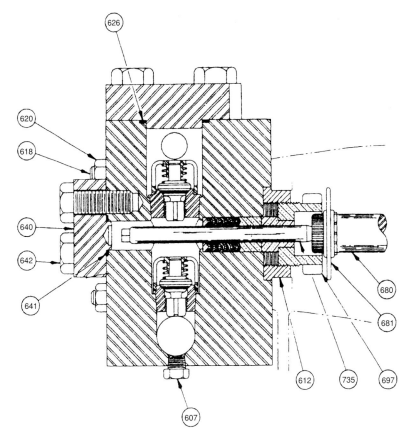

Figure 3.7. Block-style fluid end with tapered valve seats

markets. The early fluid blocks used by most old-line pump manufacturers gave way to stainless steel and exotic metals. Maybe the first step toward this change involved using a material like 4140 alloy steel and then heat-treated stainless steels (e.g., 17.4 pH, 15.5 pH, or *Custom 450*) and duplex metals (e.g., 2205 and 18–8 Mo) for the fluid block, fluid cylinders, and stuffing boxes. For special services, a corrosion data table (see Appendix A) rates some fluid-end material with pumping applications. The highest corrosion-resistant ratings for fluid-end materials when pumping certain fluids are listed here as examples:

- Alloy 20 has an *A* or best rating for salt but a *B* rating for seawater.
- The 316 stainless steel also has a *B* rating for seawater.

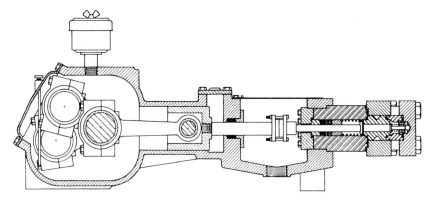

Figure 3.8. Fluid end with in-line valves developed by Aqua Dyne

- The 316 stainless steel and Alloy 20 have a *B* rating for hydrogen sulfide.
- The 316 stainless steel and Alloy 20 both have an *A* rating for methanol.

3.3   High-Pressure Packing

High-pressure pump packing also evolved from the early rope packing, square-braided packing, and cotton duct construction. Utex Industries offers a pliable packing material called *U-Pak* that has proved valuable in certain applications. Figure 3.9 shows a *U-Pak* arrangement used in a high-pressure pump. In this figure, the throat bushing (item 1) and the lantern ring (item 5) are shown to be used in conjunction with two packing rings (item 2) that face each other to contain *U-Pak* material (item 3). The insert fitting (item 6) is used to inject and pressurize the *U-Pak* material.

In a paper presented in 1998, Fred Pippert explained that in 1964, Utex Industries developed the first nonadjustable plunger packing material designed to address reciprocating pump-sealing problems. This material was composed of nitrile rubber and nylon fabric composite laminated material and was then molded into packing called the *J-Design 838*. In 1992, Utex began to investigate new elastomer systems that could allow the production of plunger packing material that could operate at higher pressures and temperatures as well as operate with less maintenance. Testing and evaluation were conducted, and in 1997, the new composite was introduced and given the name *SuperGold*.

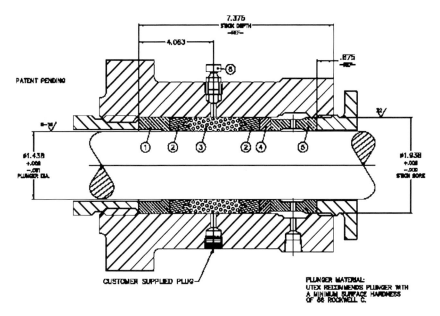

Figure 3.9. U-Pak arrangement developed by Utex

Materials such as ultra-high-molecular-weight (UHMW) polytetra-fluoroethylene, polyetheretherketone (Peek), polyphenylenesulfide (PPS), and Delrin have increased the packing life for high-pressure pumps in the water-jetting industry. Specialty packing is being used in services such as methanol pumping. Figure 3.10 shows the UHMW material used for up to 20,000-psi pump operation. In addition, a packing cartridge has been developed for methanol pumping with pressures up to 15,000 psi.

Figure 3.10. High-pressure packing arrangement developed by Reliable

3.4   Hydro-Balanced Packing

A technology called the *Hydro-Balanced (H-B) Packing System*, patented by Harold Palmour, was developed for multiplex plunger pumps in the 1990s. The standard packing arrangement shown in Fig. 3.11 can be replaced with the H-B system shown in Fig. 3.12.

The H-B stuffing box arrangement can be used in several ways as outlined in alternatives A and B, shown in Fig. 3.13. **Alternative A** is an H-B packing system consisting of a stuffing box with a "piston" as the secondary seal and V-ring packing as the primary seal. A barrier fluid is supplied under pressure to the space between the two seals so that the primary packing sees only the barrier fluid. Additional hardware is needed, such as a check valve, barrier fluid reservoir, and a compressed air source. **Alternative B** is the same H-B packing system as in alternative A, but the system is fitted with a *Button Head Ball Check* into which a barrier material can be injected in the space between the two seals. The barrier material can be high-temperature grease, *Jet-Lube*, *U-Pak* material, or a special sealing material. Additional hardware would consist of an injection gun to supply the barrier material. This alternative uses the same internal parts as alternative A, but it does not have the barrier fluid reservoir to resupply the barrier fluid automatically. In addition, there exists a third alternative (not shown in Fig. 3.13) that involves V-ring packing and could be used if a user prefers a conventional approach.

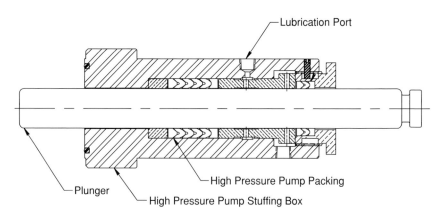

Figure 3.11. A standard packing arrangement used in Kobe pumps

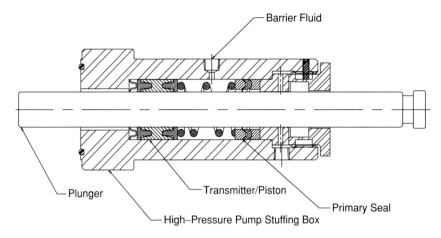

Figure 3.12. Hydro-Balanced packing arrangement tested in Kobe pumps

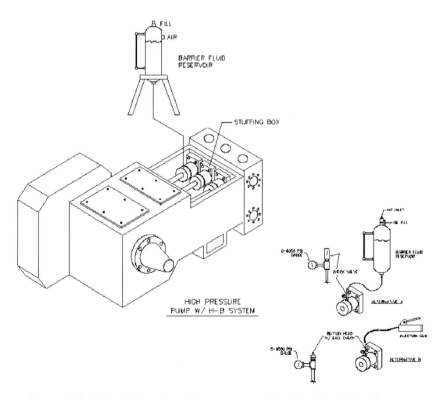

Figure 3.13. Hydro-Balanced packing arrangement developed for National pumps

3.5   The Intensifier

The company Jet Edge describes the intensifier as an amplifier because it converts the energy from the low-pressure hydraulic fluid into UHP water. The hydraulic system provides fluid power to a reciprocating piston in the intensifier's center section. A limit switch is located at each end of the piston that generates pressure in both directions. As one side of the intensifier is in the inlet stroke, the opposite side is generating UHP output. During the plunger inlet stroke, filtered water enters the high-pressure cylinder through the check valve assembly. After the plunger reverses direction, the water is compressed and exits into a pressure vessel or attenuator. The attenuator smoothes pressure fluctuations from the intensifier and delivers a constant and steady stream of water to the cutting or cleaning tool, which creates the ultra-high-pressure feature.

3.6   High-Pressure Plumbing

High-pressure fittings developed as a result of the growing demands of high-pressure pumps and systems. In the 1970s, it was difficult to find a pipe fitting rated more than 6,000 psi, and, when available, the 10,000-psi fittings were forged-steel types that were heavy and bulky. In the late 1970s, water-blasting manufacturers made pipe fittings from stainless steel bar stock and replaced the forged steel predecessors. By using stainless steel schedule 160 pipe and custom-made stainless steel couplers, a barrel could be made for the 10,000-psi handguns. Other gun manufactures used stainless steel tubing and welded fittings on each end for different gun barrel lengths. As the pressures moved to 20,000 psi, pipe fittings gave way to *Autoclave*-type metal-to-metal cone connections and tubing. BuTech, HIP, and other high-pressure fittings and tubing became available in the 1980s for the 20,000-psi, 30,000-psi, 36,000-psi, and the 40,000-psi pump systems being manufactured and installed in various applications. In more recent times, such as in the twenty-first century, 100,000–160,000-psi fittings have been developed for use in the UHP pumping industry.

3.7   High-Pressure Hose

High-pressure hose and quick disconnects also developed as the demand for high-pressure fittings and pump systems increased. In the 1960s and 1970s, a hydraulic hose was pressed into service as a water blast hose. The hydraulic hose rated for hot oil service was re-rated for "open-ended service" at 10,000 psi by companies such as Gates, Imperial Eastman, Pirelli, and AeroQuip. Large quantities of 1/2-inch I.D. hose have been used for water-blasting services, and this type of hose may be one of the main expendables in the industry. For example, a company could purchase bulk hose material and end fittings for its water-blasting service fleet. The material could then be cut to length, assembled with 1/2-inch NPT ends, and tested up to pressures of 15,000 psi before being used in the field. As time passed, the labor to do the assembly, the amount of repair needed, and the liability to the company would exceed the product's benefits, so outside vendors were used exclusively to replace the hose assembly operation. Powertrack International, based in Pittsburgh, PA, marketed a water-blasting hose for use in the range of 10,000–18,000 psi. Mike Conroy opened a branch office for Powertrack in Houston, TX, in 1984.

Until the 1980s, small hoses with fairly low-pressure ratings were used in tube-cleaning operations called *flex lancing*. Self-propelled nozzles were being used on a small black hose made by Parker to do tube lancing for pressures of up to 6000 psi; however, the hose was only rated for 4000 psi. A company called Triplex assembled several sizes of Teflon cored hose (made by a company named Teleflex) with a cover of braided stainless steel and stainless steel fittings. The hose was flexible and could easily be maneuvered in turns and bends in the tubes being cleaned. Even though this hose was rated for use with pressures much lower than 10,000 psi, companies still used it for tasks that required 10,000 psi of pressure because there was nothing else available with these features.

Synflex manufactured various sizes of hose material, and some were just right for many of the tube lancing jobs. One company stocked as much as 4,000 linear feet for its service fleet and assembled the flex lances in any length needed. Synflex decided that the hose being used for tasks requiring 10,000 psi, a hose that fit perfectly inside a heat exchanger tube, had to be increased in outer diameter, which then made it useless for the purpose it served. The service company stopped using that hose brand after the inventory was depleted and looked for another alternative. A company named Rogan & Shanley offered a braided hose that had a thermoplastic outer cover. Rogan became acquainted with his supplier, named Polyflex (based in Germany), through his research at the Imperial College in London, England. However, the hose was stiffer than hoses made with the Teleflex material, and the operators in the field easily broke the fittings. Rogan listened to the feedback from the users, and the fittings were redesigned to handle the mechanical damage caused by connecting the hose to nozzles and connectors. The newer hose soon found competition from another hose created by Spir Star (based in Germany), so the high-pressure flex lancing began to catch up to the high-pressure and UHP pumping systems. As of this writing, Tony Bessette is vice-president of Spir Star in Houston, TX. Before joining Spir Star in 1995, Bessette worked for Weatherford, Butterworth, and Rogan & Shanley.

By the 1980s, Powertrack International was well known for the creation of a water-blasting hose used for jobs requiring pressures in the range of 10,000–18,000 psi. Parker Hannifin (the firm that purchased Rogan & Shanley) and Spir Star are presently competing in the UHP hose market.

3.8 High-Pressure Accessories

Accessories for the high-pressure pumps include the "quick disconnect" or hose coupling. Some of the first accessories available for the water-blasting industry were the AeroQuip connectors used in hydraulic service. The quick disconnects made of steel were coated with cadmium and had small ball bearings to hold the two halves together. To improve the design, some of the water-blasting manufacturers began to make their own stainless steel versions of the quick disconnect. American Aero made one with ACME screw threads for quick makeup and was rated for 40,000-psi burst pressure. Other manufacturers included a metal-to-metal cone seat, and most of them could be assembled by hand to connect hose or accessories for operation at 10,000-psi working pressure.

3.9 The Radial Pump

3.9.1 Radial Diaphragm Arrangement

The *Harben Radial Piston Diaphragm Pump* is used in water-blasting and cleaning operations, including those of the runway cleaner equipment described in Chapter 12. The *Century Pump* comes in a four-cylinder and eight-cylinder design for flows to 40 gpm and pressures to 10,000 psi. The radial diaphragm arrangement allows increased flows as the pressure goes down, and it can run dry without damage as well as can handle any fluid compatible with the wetted materials.

3.9.2 Hydroblaster, Inc.

In 1965, Alexander L. Vincze founded Hydroblaster, Inc. in Sparks, NV. In 1980, Vincze applied for a patent for a constant pressure pump with axial pistons in a compact design. The company created a six-piston and eight-piston model that delivered pressures up to 10,000 psi at 4.5 gpm and 6.0 gpm, respectively. Advertising for this model was heavy during the 1980s but seemed to disappear after Vincze's death.

3.10 High-Speed Pumps

Older model pumps ran at 150–500 rpm as a maximum speed. As discussed earlier, smaller pumps and municipal pumps can run up to 650 rpm, but most of the high-pressure pumps were run under 450 rpm for longer power end life. At the extreme, the Flow International pump operates at a full driver speed of 1,800–2,000 rpm.

Starting in 1995, New Jet Technologies, based in Seattle, WA, searched for a crankcase/power end to use in a UHP pump to replace intensifier models. By 1996–1997, the *Hughes Pump* crankcase was selected, and a fluid end using various technologies was built. Pump models in the 125-hp range were developed for 4.2 gpm at 40,000 psi, as shown in Fig. 3.14. In addition, models such as those in the 185-hp range for 6.5 gpm at 40,000 psi, as shown in Fig. 3.15, were also created. The small, lightweight power end has a gear reduction of 2.85:1.0 and can run up to 750 rpm with a high crank load. The power end can bolt directly to an engine for better alignment and a smaller footprint. The pressurized lubrication system used on this power end extends the life expectancy of bearings and seals by providing cooling and filtering of crankcase oil.

3.11 New UHP Pumps

Around 1985, a company in Taiwan called OHPrecision began to manufacturer high-pressure pump spare parts that meet or exceed the standards of major OEMs, such as FLOW, I-R, and OMAX. In addition to spare parts, the company has introduced UHP pumps that take advantage of the staff's knowledge and experience in high-pressure component manufacturing. Figure 3.16 shows a 160-hp (117-kW) pump model.

Model UH-160 is a 40,000-psi (2750-bar) direct-drive pump for up to 6.1 gpm (23 lpm) at an input speed of 1800 rpm. With the built-in gear

Figure 3.14. Ultra-high-pressure pump unit (*Courtesy of New Jet Technologies*)

Figure 3.15. Ultra-high-pressure pump with Hughes power end
(*Courtesy of New Jet Technologies*)

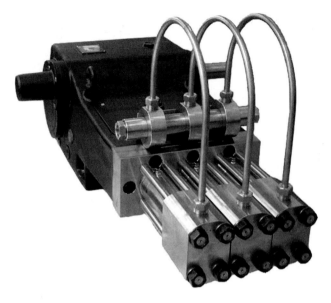

Figure 3.16. Ultra-high-pressure pump (*Courtesy of OH Precision*)

reduction of 3.389:1.0, the pump speed is 530 rpm. The triplex pump has hipped ceramic plungers and weighs 805 lb (365 kg) and can be used for water jet cleaning, surface preparation, concrete demolition, or hydrostatic testing. QualJet, located in Bellevue, WA, is a partially owned subsidiary and worldwide distributor of OHPrecision.

3.12 Contributors to Pump Evolvement

Those that have contributed to the evolvement of the high-pressure pumps include the companies in this chapter as well as those described in Chapter 1. Partek, American Aero, NLB, Aqua-Dyne, Hammelmann, Woma, Wepuko, Sugino, Kamat, Uraca, Hydro-Services, Jetstream, Jetech, FLOW, and Jet Edge are some of the companies that have developed high-pressure pumps and systems through research, innovation, and experience.

References

Gracey, M. T., and Palmour, H. H. (2000). Artificial lift hydraulic pump improvement. *Southwestern Petroleum Short Course*, Lubbock, Texas.

Pippert, F. (1998). Elastomer technology results in improved performance while requiring less lubrication. *Oilfield Engineering with Polymers Conference II*, London, England.

Vincze, A. L. *Constant pressure pump*. United States Patent, October 11, 1988, number 4,776,260.

Chapter 4

Development of Nozzles and Accessories

4.0   High-Pressure Accessories

The development of nozzles and accessories for pumps sometimes occurs after high-pressure pump improvements and sometimes is closely related to pump capability development. Accessories must be developed to accommodate higher pressures when needed. Accessories and nozzle technology has also improved the effectiveness and usability of high-pressure pumps in water-blasting and water-jetting applications.

4.1 Water-Blasting Guns

High-pressure, positive displacement pumps were sometimes used in the oil field industry as early as the 1930s. In the 1950s, operators of waterflood trucks had the idea their vehicles with water at the pressure of 500–1,500 psi, which was produced by the positive displacement triplex pumps mounted on board. The end of a piece of pipe was flattened to form a fan spray nozzle for the washdown operation. In the late 1950s, brine truck and well-kill truck operators used the 2,000- to 3,000-psi triplex pumps to blast away mud and salt from their trucks after deliveries into remote areas of the oil patch. It was reported that a pipe cap was drilled to make a nozzle, which was used on a piece of pipe as a cleaning wand. This may have been one of the first water-blasting guns invented. The gunman would hold the piece of pipe and signal the operator to engage or disengage the pump for the washdown operation because there was no way for him to control the flow to the nozzle.

The next improvement to this water-blasting invention may have been a handle on the piece of pipe to help the gunman control the cleaning wand; around 1963, a needle valve was added to the wand to allow the water to be dumped out instead of going through the nozzle— this provided a measure of safety and control for the gunman. By 1963, the wand was refined with pipe fittings, a manual dump valve, and a replaceable nozzle. With the development of pump packages designed for water blasting, KACY Manufacturing probably offered one of the earliest 10,000-psi dump-style water-blasting guns on the market. It was hard to hold the trigger because of the back-pressure of the water, so some users welded extensions on the trigger to help make it usable. CCI Pumps also had a version of the dump gun that reportedly did not work well. In 1967, Clem Mondy of Job Master devised a dump-style gun because there did not seem to be one that was suitable for marketing with his water-blasting equipment. This gun proved to be lighter in weight, more reliable to use, and easier to hold than other water-blasting wands. It undoubtedly became the model for several manufacturers in the evolving water-blasting industry. For example, Jack Hinrichs of Hydro-Services patented an air control system in 1968 based on this model, as shown in Fig. 4.1.

Hinrichs also patented an improvement to the dump-style gun in 1972; he created a double cam-action trigger to make the gun easier to operate, which reduced gunman fatigue. These ideas developed into a number of gun styles and innovations, such as shut-off-style guns, unloader valves, remote-control type systems, and, eventually,

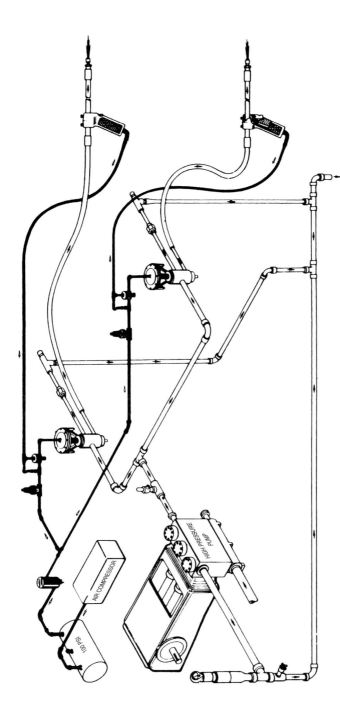

Figure 4.1. Air control system for multiple gun operation (*Courtesy of Hydro-Services*)

ultra-high-pressure (UHP) guns. Examples of the early gun types are shown in Fig. 4.2.

The dump-style gun relieves the pressure in the system until the trigger is operated; for this reason, the gun has been called a *deadman valve*. The shut-off gun normally holds pressure in the system and requires an unloader or pressure regulator at the pump to bypass the flow from the pump. The air control gun has a small ball valve at the trigger that operates a poppet valve mounted on a two-way diaphragm valve so that the water is not furnished to the gun until the trigger is operated. The variations of the air control system include multiple gun operations and multiple stations. As water-blasting operations increased in scope, there was a need for foot-operated guns to control rigid and flexible lances used to clean heat exchangers and pipes (foot guns are typically similar to companion hand-operated guns). If a dump-style handgun were used, a dump-style foot gun would also be used when needed. Figure 4.3 shows various early foot gun configurations.

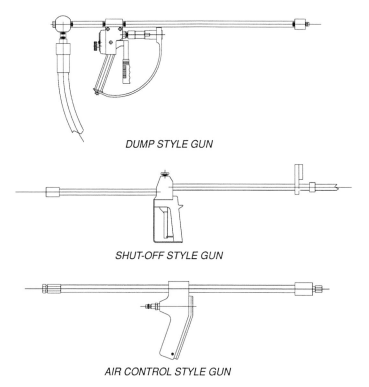

DUMP STYLE GUN

SHUT-OFF STYLE GUN

AIR CONTROL STYLE GUN

Figure 4.2. Examples of early water-blast handgun types

Once the water-blasting pumps overcame the 6,000-psi barrier, companies set out to reach the next plateau of 10,000 psi. Much of the equipment, including the positive displacement plunger pump, was set up for 10,000 psi by the mid- to late 1970s. It became a rule of thumb that a gunman could hold about 10,000 psi at 10 gpm. This rule, of course, varies greatly with the size of the gun-man and the position of the operator. Figure 4.4 shows the recommended maximum operating pressure by body weight and nozzle orifice size as calculated by the following formula:

Pressure = body weight divided by 4.71 and multiplied by the nozzle diameter squared

$$\Delta P = \frac{\text{wt.}}{4.71}\left(\frac{\text{nozzle}}{\text{DIA}}\right)^2$$

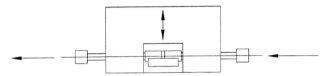

FOOT OPERATED SHUT-OFF STYLE GUN

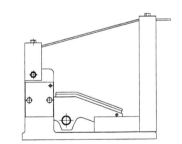

FOOT OPERATED AIR CONTROL GUN

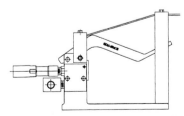

FOOT OPERATED DUMP GUN

Figure 4.3. Examples of early water-blast foot gun types

Recommended Maximum Operating Pressure
(By Body Weight And Nozzle Diameter)

D I A M E T E R (Inches)

		1/32" 0.031	0.038	0.041	3/64" 0.047	0.054	1/16" 0.063	5/64" 0.078	3/32" 0.094	7/64" 0.109	1/8" 0.125
B	160	35,349	23,525	20,208	15,378	11,650	8,559	5,584	3,845	2,859	2,174
	165	36,454	24,260	20,840	15,859	12,014	8,826	5,758	3,965	2,949	2,242
O	170	37,558	24,995	21,471	16,339	12,378	9,094	5,933	4,085	3,038	2,310
	175	38,663	25,731	22,103	16,820	12,742	9,361	6,107	4,205	3,127	2,378
D	180	39,767	26,466	22,734	17,300	13,106	9,629	6,281	4,325	3,217	2,446
	185	40,872	27,201	23,366	17,781	13,470	9,896	6,456	4,445	3,306	2,514
Y	190	41,977	27,936	23,997	18,262	13,834	10,164	6,630	4,565	3,395	2,582
	195	43,081	28,671	24,629	18,742	14,198	10,431	6,805	4,686	3,485	2,650
	200	44,186	29,406	25,260	19,223	14,562	10,699	6,979	4,806	3,574	2,718
W	205	45,291	30,142	25,892	19,703	14,926	10,966	7,154	4,926	3,663	2,786
	210	46,395	30,877	26,523	20,184	15,290	11,234	7,328	5,046	3,753	2,854
E	215	47,500	31,612	27,155	20,664	15,654	11,501	7,503	5,166	3,842	2,921
	220	48,605	32,347	27,787	21,145	16,018	11,768	7,677	5,286	3,931	2,989
I	225	49,709	33,082	28,418	21,625	16,382	12,036	7,852	5,406	4,021	3,057
	230	50,814	33,817	29,050	22,106	16,746	12,303	8,026	5,527	4,110	3,125
G	235	51,919	34,553	29,681	22,587	17,110	12,571	8,201	5,647	4,199	3,193
	240	53,023	35,288	30,313	23,067	17,474	12,838	8,375	5,767	4,289	3,261
H	245	54,128	36,023	30,944	23,548	17,838	13,106	8,550	5,887	4,378	3,329
	250	55,233	36,758	31,576	24,028	18,203	13,373	8,724	6,007	4,468	3,397
T	255	56,337	37,493	32,207	24,509	18,567	13,641	8,899	6,127	4,557	3,465
	260	57,442	38,228	32,839	24,989	18,931	13,908	9,073	6,247	4,646	3,533
(lbs)	265	58,547	38,963	33,470	25,470	19,295	14,176	9,248	6,368	4,736	3,601

$$\text{PRESSURE} = \frac{\text{BODY WEIGHT}}{4.71 \times \text{DIAMETER SQUARED}}$$

Figure 4.4. Table of recommended max. operating pressure by body weight

The nozzle diameter relates to the flow produced at a given pressure, and, therefore, it is proportional to the thrust that the gunperson experiences. The major component of the thrust is the flow as indicated in the following formula:

Thrust = 0.05260 × flow × square root of pressure

$$T = .05260 \times Q\sqrt{\Delta P} = .0526 \times \text{gpm}\sqrt{\text{psi}}$$

During the next 20 years, water-blast guns did not experience much change except for the addition of operator safety features, such as trigger guards, safety catches, safety hose shrouds, and secondary dumps. One of the first safety features to appear on the water-blasting gun was a guard around the trigger. Shut-off guns used in pressure washing were probably the first to add safety catches, but they did not appear on the high-pressure guns until the 1980s. Some guns were made with a high-pressure hose near the operator, so a hose safety shroud was developed in the 1970s and continued to improve in design over the years. At first, a piece of fire hose was used to create the shroud, and then materials such as Kevlar covered with a slick coating were developed to shield the gunman from a leak or rupture in the high-pressure hose. Because a gun mechanism can fail to operate properly, the secondary dump and the two-hand water-blasting gun became mandatory in some petro-chemical plants. The first known solution to meet the requirement was the connection of two dump-style guns; in case one did not work, the other was expected to operate correctly. Dirt, worn parts, or some sort of jam or obstruction could also cause a malfunction in a manually operated gun. Secondary dump features were incorporated into all of the handheld water jetting guns as a backup to the primary trigger to meet the safety requirement where enforced. In addition, there have been gun part improvements, such as the creation of the replaceable valve cartridge introduced by Jetstream in late 1982, as well as continuing improvements in seals and materials. Internal gun parts have been under continual scrutiny for refinements in their operation and dependability.

Manufacturers and equipment users are very interested in safe water-jetting operations because the personal injury and legal costs are too high to take a chance with safety. When 20,000 psi became the next pressure plateau, the handgun and foot gun were converted to autoclave fittings but had little internal part changes. The next development in high-pressure water jetting was UHP guns to go with the intensifier pumps that operated at 36,000 psi. Flow Industries probably had the first 36,000-psi handheld guns, and Jet Edge developed a wand around 1980 called the Gyrajet that included a hydraulically powered rotating nozzle and a gun called *Litelance* that featured electric actuation of a remote-control valve for 36,000-psi operation. In the period from 1983 to 1987, Michael Mullinaux worked as a sales/service manager for ADMAC, during which time Jetlance continually improved. One of the company's customers was Gene Valentine of Valley Systems of Ohio, and Valentine developed a tool called the *Orbital Jettol*. The tool was successful in Valley Systems's business of servicing Dupont plants. Valley Systems, in fact, was one of the pioneers in industrial maintenance and continues to function in the twenty-first century as part of HydroChem. High-pressure and UHP guns also continue to make more use of hydraulics, pneumatics, servomotors, pilot valves, and electronics to increase reliability, decrease fatigue, and decrease equipment weight.

4.2   High-Pressure Nozzles

Special nozzle technology has been developed, tested, and patented over the years. Around the 1960s, a mechanical device was used to break up a water jet into a series of water slugs that created a sequence of water-hammer impact stresses on a surface. The low-frequency mechanical device had drawbacks because of internal erosion and wear, but the principle was further developed by Hydronautics in Laurel, MD, in the 1970s. The National Maritime Research Center in Galveston, TX, funded a research project using Hydronautics's technology to clean marine growth from the underside of ships in 1974. The positive displacement pump used in the tests was in the pressure range of only 2,000 psi and could deliver 100 gpm, but the test results proved that the pulsed nozzle could effectively remove a heavy buildup of marine growth. Hydronautics continued to develop nozzle technology under the registered trademarks of *Cavijet*, *Servojet*, and *Stratojet*.

4.2.1   The *Servojet*

The *Servojet* is a self-resonating, pulsed water jet nozzle with no moving parts that creates slugs of water to increase the rates of cleaning or cutting with high-pressure water. The *Servojet* enhanced cleaning and cutting rates using these features:

- Larger impact stresses due to the water-hammer pressure that enhances the local erosive intensity
- Larger outflow velocities across the surface being cut or cleaned, thus providing an added mechanism for opening cracks and flaws in a material or enhancing the washing action during cleaning operations
- Greater ratio of impacted area-to-volume of jetted water, thus exposing larger areas of the surface to the water-hammer pressure
- Cycling the loading, which promotes unloading stresses, thus enhancing the process of debonding the substance from the substrate or fracturing the target material
- Short duration loading, which tends to minimize energy losses in the substrate being cleaned or the material being cut and, therefore, increases the work being done for the amount of energy expended

4.2.2 The *Cavijet*

The *Cavijet* consists of a center body placed in a precise position in a nozzle orifice to produce a low-pressure area. This device is one of the very few successful attempts to harness the destructive power of cavitation for useful purposes. Virgil Johnson watched the propeller tests being done at Hydronautics and wondered how the destructive force of cavitation could be designed into a nozzle. Because cavitation on the propellers was observed under water, a large model was built and tested in a water tank controlled by a variable-speed nozzle carriage mechanism. The tests were impressive and opened the door for research projects using the technology concept. Johnson had the idea that the same technology, specifically a pulsed water-jetting nozzle, could be used to cut grass and water the lawn at the same time. Even though Johnson did not get to build his cavitating lawnmower, a prototype lawnmower-type device was later designed to inject fertilizer into the soil using the nozzle technology. The basic concept of a cavitating nozzle design is the induction of the explosive growth of vapor-filled cavities in a stream of high-pressure fluid that will violently collapse on or near the target surface. By proper adjustment of the distance between the nozzle and the surface, the cavities will implode in the high-pressure stagnation region where the jet impacts the solid material. At the same pressure and flow (horsepower), the cavitating jet provides a great advantage over noncavitating jets because it produces extremely high-localized stresses over many very small areas of the target material. Figure 4.5 shows a center body *Cavijet* nozzle made for cutting materials.

4.2.3 The *Statojet*

The *Statojet* (structured, self-resonating *Cavijet*) was first developed for an oil drill bit that operated in high-pressure down-hole conditions. In a pressure chamber, it was determined that the pulsed nozzle design did not perform in the high-pressure atmosphere, so an organ pipe design was developed to pulse even under the high-pressure situation. It was found that the design also worked well when the work could be submerged in the cleaning fluid. The concept was used to clean drill pipe by Weatherford throughout the 1980s and was incorporated into demilitarization equipment around 1986, as discussed in Chapters 5 and 6. The Weatherford tubular cleaning system that used the *Stratojet* nozzle could clean drill pipe that was completely plugged with hard

Figure 4.5. Center body cavitating nozzle

cement; for example, the nozzle could clean plugged 2.44-I.D. × 33-feet long tubulars in less than 8 min, with a best time of under 5 minutes. Figure 4.6 shows a pulsating nozzle for cleaning tubular shapes.

As a side note, all nozzles tend to cavitate under water, so some companies have attempted to sell their "cavitating" nozzles by demonstrating how they work under water because the nozzles would not cavitate in the air. In the pressure-washing industry, the closest thing to the "work" is the nozzle. The nozzle can exist in many forms, such as a straight jet, fan-jet, or shaped jet. As the beginning of this chapter

Figure 4.6. Pulsating nozzle for tube and pipe cleaning

explained, operators of water-flood trucks in the 1950s washed their trucks that had just been in the oil fields with the 500–1,500 psi produced by a positive-displacement pumps that were mounted on board. The end of a piece of pipe was flattened to form a fan spray nozzle for the washdown operation. Later, a drilled hole in a pipe plug made a straight jet nozzle. These developments certainly did not mark the invention of nozzles or even water-jetting guns, but these events occurred because of necessity.

The basic design of a simple nozzle is that it restricts the flow from a positive displacement pump to exchange pressure for velocity. If a pump can deliver 5 gpm and handle 3000 psi, it would need a 10-hp driver that could be determined from this formula:

Horse power $= 5\,gpm \times 3000\,psi \div 1714$ *(hydraulic constant)* $= 8.75\,hp$

$$HP = \frac{GPM \times psi}{1714}$$

A pressure washer has a 10-hp engine because there is a volumetric efficiency and a mechanic efficiency (or deficiency) for the pump and driver; therefore, the manufacturer may have used an 85% efficiency to cover the deficiencies. These deficiencies also account for users choosing the number 1540 instead of 1714 for the hydraulic constant when performing calculations. A nozzle for 5 gpm and 3,000 psi could have a single hole or orifice and attach to the end of a wand or pressure washer gun. The size of the orifice can be determined with the following formula:

$$\varnothing = .2046\sqrt{\frac{Q}{n\sqrt{\Delta P}}}$$

If the formula for area were used, the 0.06-in diameter would calculate to about 0.00282 in^2. The area needs to be the same for a straight jet or a fan-jet to produce the same pressure. If the nozzle has one straight hole, then a 0.06-in diameter drill will produce 3000 psi at 5 gpm. Relying on these numbers can be dangerous, however, and many pressure washers do not have pressure gauges because of this reason. Just because the horsepower, pump size, and 3000 psi are written on a pressure washer's packaging, a user should not totally rely on this information—a pressure drop can change the parameters. For example, if a restrictive hose (long and/or small) is run from the pump to the gun and nozzle, the pressure reading at the nozzle will not be the same as the pressure reading at the pump. One quick way to see the pressure drop in a system is to place a gauge at the discharge of the pump and remove the nozzle from the system. The pressure you read on the gauge is the pressure drop or "back pressure" of the system. The more water you try to push through that system, the more pressure drop you will get. An executive for a water-blasting contractor once stated that hose length did not matter and if the pump can produce the water, the hose can take it. Two things convinced him otherwise: (1) a new computer program considered the elements of the system, such as hose size and length, valves in the system, and pump capacity, to design multiple orifice nozzles; and (2) a pump unit was outfitted with hose, foot gun, and flex lance with a pressure gauge at the pump and another at the end of the hose. He saw the difference, but it was more convincing to remove the nozzle and read the pressure gauge at the pump. Another example concerning the importance of understanding how a pressure system works is found in a user's complaint about a new foot gun that he just purchased. When he operated the foot gun, the engine would die on his 10,000-psi, 10-gpm water-blasting pump unit. It took a long discussion with the manufacturer of the gun to determine that he had attached a very small rigid lance in the system and that the pressure drop was about 9,000 psi when the gun was operated. The user had not understood that 10-gpm of water would not go through the small I.D. pipe.

4.3   Back Thrust

The thrust that is felt when a gun is triggered is caused by the jetting action of the water, but the main component of the thrust formula is the water flow. An operator may hold both a gun that is running at 4,000 psi and one that is running at 40,000 psi because the thrust primarily depends on the volume of water being pumped. The 40,000-psi pump may deliver 5 gpm, so the thrust would be about 52 lb; the 4,000-psi gun may deliver 20 gpm, so the thrust could be 66 lb, as shown in the following formula (as discussed in Section 4.1 and shown in Table 4.4):

$$THRUST = .05266 \, Q\sqrt{P}$$

Understanding the thrust generated by flow through a nozzle might help a sewer jetter or water blaster determine the amount of pull for a line mole or for holding a wand or gun. For example, the sewer nozzle will need more thrust for pulling (orifices pointing back) than any orifices that point forward, while an operator of a handheld nozzle might not be heavy enough to hold the gun for very long. To calculate the thrust on a tube nozzle with angled orifices, vector algebra can be used to determine the pull generated by the nozzle.

4.4 Nozzle Devices

A "lawnmower" or floor cleaner, as shown in Fig. 4.7, is supposed to cover larger areas and do a better job than a handheld wand. If you have seen the machine leave "stripes" or clean unevenly, here are the reasons. A straight nozzle at the end of a rotating arm will clean a small path. A fan nozzle will clean a wider path, but it is less concentrated on the work surface; therefore, this type of nozzle provides less cutting action. Fan nozzles can also be more intense near the center and less intense at the edges, so uneven cleaning occurs. The rotational velocity of the nozzle and the movement rate of the nozzle over the surface must be considered (Fig. 4.8). The synchronizing of the rotation nozzles and the movement rate can be determined with the following formula:

Path width × rotational speed = optimal movement over the surface.

The rotational speed for any given job is based on the type of material being removed and is generally determined by testing under controlled conditions. Once the effective nozzle velocity along with the pressure and flow needed to do the job are determined, adding nozzles will allow the movement over the surface to be increased. The formula

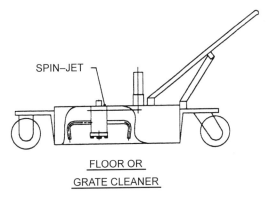

SPIN–JET

FLOOR OR
GRATE CLEANER

Figure 4.7. Lawnmower-type floor grate cleaner

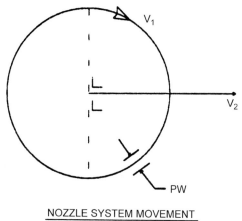

NOZZLE SYSTEM MOVEMENT

Figure 4.8. Nozzle system movement diagram

used to determine the expected speed of movement over the surface then becomes:

$$V_2 = nPw \times RPM \ (PW = \text{path width, n = number of nozzles})$$

This concept could be explored further, but it suffices to say that there are limits to the rotation speed and the forward movement; when nozzles are added, horsepower must also be added to the system to maintain the cleaning effectiveness of the increased travel speed. The topic of rotating nozzle technology is discussed later.

4.4.1   Rotating Nozzles

Several types of rotating nozzles are used for pressure washing and water blasting. A self-rotating or power-rotated head on the end of a wand or gun has been shown to be very effective to increase coverage and cleaning efficiency. Three-dimensional rotating nozzles are used for tank cleaning, and two-dimensional nozzles are used for pipe and tower cleaning. The principles used in the self-rotating nozzles rely, on part of the thrust of the water jet to cause rotation or spinning of the orifices. Using the thrust formula, it can be determined how much force the nozzle will exert in one direction. Using vector algebra, the amount of rotational thrust can be controlled by the angle that the water jet strikes the surface to be cleaned. The amount of turning movement generated by the thrust will be determined by the offset (length of the nozzle arm).

In some cases, self-rotating action has reduced the effectiveness of the nozzles. For example, a water-blasting contractor had a job removing coatings from a parking garage floor and purchased a lawnmower-type floor and grate cleaner. The material was thick urethane and required 20,000 psi to cut through it. The floor cleaner had a self-rotating design and cut the material from the floor, but the cleaner did not accomplish the task fast enough to finish the job on schedule or within the price quoted. This self-rotating floor cleaner was modified to include a controlled hydraulic rotation of the nozzles so that they could be pointed in the direction of rotation—the results were gratifying. The self-rotating design removed about 200 ft^2 of material per hour, and the power rotation design removed about 600 ft^2 per hour.

4.4.2 Specialty Nozzles

Special nozzles are often used on pressure washers; these nozzles include the adjustable straight-to-fan pattern nozzle, the rollover nozzle (that can quickly switch from one nozzle to the other), and certain nozzles that are not sold on the open market. Examples of specialty items are the pulsed nozzles and cavitating nozzles that were developed in the 1980s and installed into systems of certain facilities, such as Three Mile Island nuclear power plant and government facilities (to remove explosives from munitions and solid propellants from missiles and rocket motors, as discussed in Chapters 5 and 6). The technology found its way into the water-blasting and pressure washer industries because of long-time relationships between companies and personnel. A nozzle that cavitates was found to cut materials with lower pressures and therefore lower horsepower. It could erode and cut material when mounted on the end of a gun or when built into mechanical systems, such as concrete cutters, metal erosion test machines, and paint stripper devices. The pulsed nozzle was easily adapted to the end of a pressure washer wand and increased the effectiveness of the standard pump unit. The pulsed nozzle consists of a resonating chamber that causes the water to pulse as it leaves the nozzle, which enhances the cleaning effect. In addition to the specialty nozzles shown in Figs. 4.5 and 4.6, there is a nozzle called a *Servojet* (described earlier) and is shown in Fig. 4.9; the *Servojet* is also described in a paper by Georges Chahine *et al.* (1983). In Fig. 4.9, item 1 is a coupling, item 2 is an orifice, item 3 is a nipple, item 4 is nozzle D1, item 5 is a spacer, item 6 is nozzle D2, and item 7 is a reducing coupler.

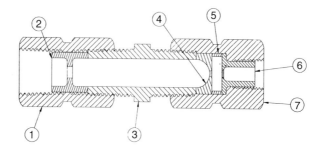

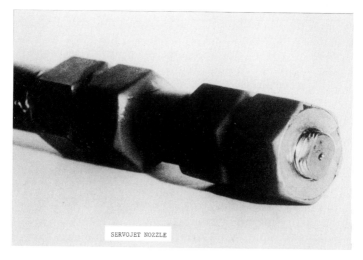

Figure 4.9. Pulsating nozzle chamber for surface cleaning

The subject of nozzles covers a wide range of industries and uses, but this discussion has attempted to mention some of the nozzles used in the pressure washing industry and for water-blasting work. A company named Spraying Systems has been one of the main suppliers of nozzles over the years, and their nozzles are used for pressures up to 10,000 psi and beyond. For work in the range of 10,000–20,000 psi, carbide-inserted "shaped" nozzles have been used that have a square- or diamond-shaped orifice. For work in the range of 30,000–40,000 psi, sapphire-inserted nozzles are common because of the increased wear factor from the water jet. There are expensive specialty nozzles on the market, and research is being done on new designs for various applications.

4.5 Multiple Gun Systems

After water-blasting guns were created, companies began developing multiple gun systems. The reasoning for their creation was that if a gunman could only hold 10,000 psi at 10 gpm (75 hp), then two gunmen should be able to operate a high-pressure pump that delivers 10,000 psi at 20 gpm (150 hp). When someone wanted to operate two guns at one time from one pump, a tee was added to the discharge line of the pump, and each nozzle was sized to deliver half of the flow at the desired pressure. If dump-style guns were used, both gunmen would have to operate at the same time, or there would be no pressure in the system. By the 1960s, orifices were added at the tee to maintain pressure in the system even with one gun operating, but it was not a satisfactory solution to the dual gun concept because of the inefficiency of the system. One manufacturer designed a gun with built-in orifice spools to act as a flow divider, but the concept was not widely accepted. The invention of the unloader for 1,000–2,000 psi used in the pressure washer industry could have been easily used for a multiple shut-off gun operation. For a dual gun operation, each gun nozzle would be sized for half of the flow at the desired pressure. Meanwhile, by the 1970s in the high-pressure water-blasting industry, two or three water-blaster manufacturers developed spring-loaded orifice mechanisms that were an improvement over the fixed orifices. By the 1980s, an adjustable flow divider was available that could balance the flow between guns fairly well for multiple gun operation. The multiple gun system patented in 1968 by Hinrichs paved the way for the invention of remotely actuated UHP wand control systems (sometimes called a *tumble box*), and the resulting power rotating nozzle guns that are currently used in the twenty-first century.

Positive displacement pumps create a certain flow at a given pump speed, so any multiple gun system works basically the same way to work smoothly. The pump must see a constant back-pressure while each handgun is working or not working. When one gun is connected to a high-pressure pump, the nozzle is usually sized to take the entire flow at the desired pressure, or the excess flow must be dumped away by an unloader or bypass valve. In a dual gun system, for example, each gun nozzle is sized to take half the flow at the desired pressure. In either case, the desired pressure will not be achieved if the nozzles in the guns are too large, or the system will overpressure if the nozzles are too small so that a bypass or an unloader valve of some sort will be required.

Floating spring-loaded orifices were the first attempt to balance two water-blasting guns—the design worked, but sometimes it worked spasmodically. One multiple gun system designed for up to 20,000 psi uses shut-off style guns and a nitrogen pressure regulator that can be adjusted to bypass at a set pressure. An improvement over the fixed orifices and the spring-loaded orifices was the adjustable multiple gun control system that uses dump-style guns balanced by spring-loaded needle valves in a common manifold. The systems on the market come in 10,000- and 20,000-psi versions. Some systems developed in Germany and Japan use an unloader valve with shut-off guns in the 10,000-psi range, but not many U.S. manufacturers use unloaders in their 10,000- to 20,000-psi high-pressure pumps systems because of the maintenance and the wide use of dump-style guns. The most dangerous aspect of multiple guns is the possibility of the gunmen blasting each other, even though this occurrence is very rare; in addition, danger lies in not balancing the system properly or the user not understanding how the system works. If a high-pressure pump produces more flow and pressure than one person can handle using one gun, a multiple gun system could be an advantage to increase the amount of work one person can do. While some owners and potential buyers assume that twice the work can be done with a multiple gun system, this is not always the case. More time is needed for setup and maintenance of a mobile system, while additional equipment has to be purchased such as guns, nozzles, hoses, and hardware. The greatest advantage of multiple gun systems might be their ability to be part of fixed or permanently installed workstations.

4.6 Nozzle Principles

Development of nozzles and accessories has increased the use of high-pressure pumps and systems for water-jetting applications. To illustrate how some people misunderstand the nozzle, a salesman once explained that he bought a straight nozzle for his garden hose, and it boosted the pressure much higher than what was provided by the city water supply (which is about 40 psi). He incorrectly thought that the concentrated round jet created a higher pressure than a regular garden nozzle. The maximum pressure in the hose can only be 40 psi if the city water is at 40 psi; a garden hose can have a blank plug on its end, and the pressure will still be 40 psi. The principle here is somewhat different for a positive displacement pump because the pump produces flow (not pressure). The pressure is developed when a nozzle, valve, or other device restricts the flow. The pressure in the hose or gun for a pressure washer is determined by orifice size (given that the pump and driver can handle it). If the flow is stopped, something is going to give. Water-blasting guns can be dump-style, so that the water goes to the atmosphere when not triggered, or can be shut-off style like most pressure washers. The unloader on a pressure washer is designed to bypass the water when the shut-off gun is not in use. The basic design of a simple nozzle is that it restricts the flow from a positive displacement pump to exchange pressure for velocity.

4.6.1 Hydrojet Nozzle

The *Hydrojet Nozzle* marketed by Hydro-Engineering is a self-rotating spin nozzle for UHP cleaning and surface preparation. As shown in Fig. 4.10, the nozzle has a 9/16 hp female autoclave connection for operation from 15,000 to 55,000 psi and a maximum flow of 15 gpm. It weighs 3 lb, and the manufacturer claims that the nozzle is reliable, is horsepower efficient, lowers costs, is comfortable to use, and is repairable.

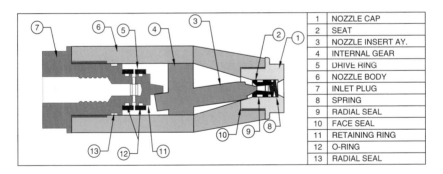

1	NOZZLE CAP
2	SEAT
3	NOZZLE INSERT AY.
4	INTERNAL GEAR
5	DRIVE RING
6	NOZZLE BODY
7	INLET PLUG
8	SPRING
9	RADIAL SEAL
10	FACE SEAL
11	RETAINING RING
12	O-RING
13	RADIAL SEAL

Figure 4.10. Hydrojet nozzle diagram (*Courtesy of Hydro-Engineering*)

4.6.2 Pulsed Jet Machine

Advanced Technologies offers the *Pulsed Jet Machine*, which consists of a high-pressure pump, ultrasonic generator, and accessories to energize the transducer, mounted on a gun or mechanism. The equipment is shown in Fig. 4.11 and includes a forced pulse generator.

The machine can be operated as a forced pulsed jet machine or as a continuous water-blasting machine by switching the ultrasonic generator on or off. The nozzle tip used at the end of the handgun is

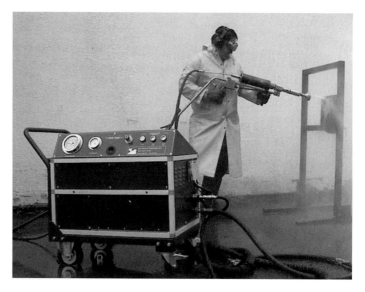

Figure 4.11. Forced pulse machine (*Courtesy of Advanced Technologies*)

diagrammed in Fig. 4.12. In an article for the *Journal of Protective Coatings & Linings*, the development and tests of the concept reveal an operation at low pressures, dependability, and a compact design for this technology.

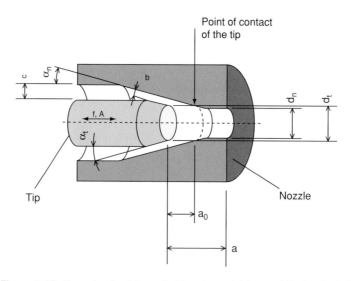

Figure 4.12. Forced pulse jet nozzle (*Courtesy of Advanced Technologies*)

Figure 4.12. Continued

4.6.3 *Sidewinder*

After New Jet Technologies focused on establishing itself as a high-pressure water-jetting service company in the early 1990s, the company started making high-speed intensifiers and simplified power units for

Figure 4.13. Ultra-high pressure rotary jet (*Courtesy of New Jet Technologies*)

Figure 4.14. (Courtesy of New Jet Technologies)

use with rotary tools. The *Sidewinder* 40,000-psi ultra pressure rotary jet tool was developed for 6 gpm at 40,000 psi, as shown in Fig. 4.13.

This pneumatic tool has been used to remove paint, coatings, and fouling at rotating speeds of up to 5000 rpm. Other accessories developed by New Jet include floor blasters, multiple valve systems, and *A-jet* cutter nozzles. Figure 4.14 shows the DTV multi–tool control valve for direct-drive or compensated UHP systems and abrasive water-jet hybrid tools.

References

Chahine, G. L., Johnson, V. E., Jr., Conn, A. F., and Frederick, G. S. (1983). Cleaning and cutting with self-resonating pulsed water jets. *Second U.S. Water Jet Conference*, Rolla, Missouri.

Gracey, M. T. (1989). Industrial applications for rotating nozzle technology. *Fifth American Water Jet Conference*, Toronto, Canada.

Yan, W., Tieu, A., and Baolin, M. V. (2003). High-frequency forced pulsed water-jet technology for the removal of coatings. *Protective Coating Lining*.

Chapter 5

Specialty Systems: The Rocket Propellant

5.0 Specialty Pump Systems

High-pressure pumps are used for special purpose systems, such as for removing the propellant from rocket motors. Similar equipment is also used to remove coatings or unwanted material in drill pipe, casing, and tubular shapes, such as industrial pipe. Oil-field tubulars and heat exchanger tubes were first cleaned by rigid and flexible lances using high-pressure water jet nozzles. Tube cleaning machines were then developed with a rotating lance on a track to improve the safety and efficiency of cleaning the internal diameter of pipe and tubes. The technology for cleaning defective rocket propellant from rocket motors may have developed independently but was definitely influenced by the tube lancing machine designs.

5.1 Rocket Propellant Washout System

A water-jetting system using a pulsing nozzle to remove propellant from rocket motors was designed and built for a group outside the United States. The special nozzle technology of the *Rocket Propellant Washout System* allowed the use of 10,000 psi instead of higher water-jet pressures. The rocket motors ranged from 0.5 meters in diameter by 4 m in length to 1 m in diameter by 6 meters in length. The goal was to build a system that could operate at pressures that were well below the level capable of detonating explosives or propellants, to provide for a man-in-the-loop to observe and control, to integrate automation and simplicity while maintaining safety, and to provide a more efficient, foam-free process. Earlier hot-water washout systems created foam when attempting to remove rocket propellant.

5.2 Nozzle Design

Previous work done with cavitating and pulsing nozzles to remove high explosives (HE) from munitions and solid propellants from missiles and rocket motors contributed to the design of a complete system to suit the customer's unique requirements. The hardware was developed to fit into a facility in Pakistan and had features that allowed the use of a standard high-pressure pump to operate at lower pressures, prevent detonation of the propellant, and integrate automation while providing operator-safe control for the whole process. The complete system consisted of a tubular washout cradle, lance mechanism, high-pressure pump unit, control console, effluent catch tank, water filter, make-up water tank, water chiller, filter transfer pump, and filter tank pump, as shown in Fig. 5.1. Figure 5.2 shows the *Rocket Propellant Washout Lance* track and high-pressure pump unit.

5.3 Safety Features

The nozzle technology for the *Rocket Propellant Washout System* came from a design used to clean oil-field tubulars while keeping the work flooded. The pulsing nozzle works best when the item, part, or tubular is in the flooded mode, and it also makes removing propellant safer. With the nozzle working in a flooded condition, it is virtually

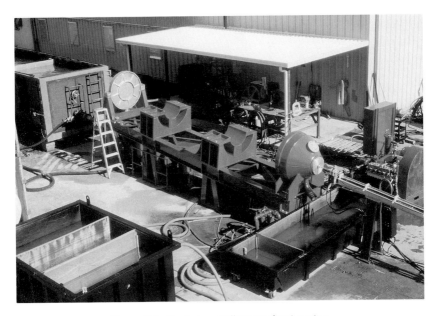

Figure 5.1. Rocket propellant washout system

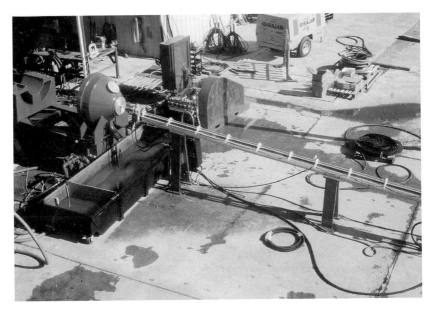

Figure 5.2. Rocket propellant washout lance and high-pressure pump unit

Figure 5.3. Pulsating nozzle assembly for rocket washout

impossible to create enough heat or a spark to set off the propellant. Figure 5.3 shows the nozzle assembly used in the system with the unusual over-the-center design.

The nozzle body design of the pulsing nozzle is shown in Fig. 5.4, and the removal inserts that help create the self-resonating feature are shown in Fig. 5.5. The family of pulsing, cavitating, and resonating nozzles is discussed more fully in Chapter 4.

5.4   Water Recirculation

The *Rocket Propellant Washout System*'s water recovery and processing was accomplished by collecting the effluent from the washout process in a catch tank and then circulating the water through a baffled filter tank and through a bank of filters before it entered a water storage tank. The baffle tank, as shown in Fig. 5.6, further removes propellant particles from the water after the catch tank has removed much of the larger pieces from the effluent.

The water recirculation system design removes virtually all of the solid material from the water so the owner/operator can collect the propellant for disposal. Another feature of the system is a water chiller to remove the heat of the jetted water before it is returned to the

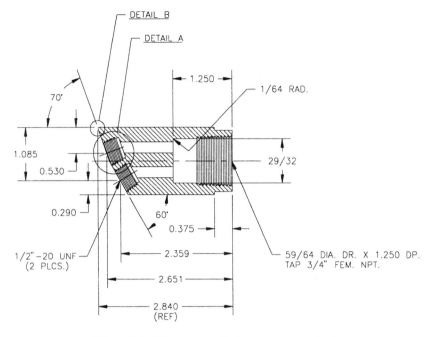

Figure 5.4. Pulsing nozzle body for rocket washout

storage tank. It has a certified capacity of 50 gpm of water from 96° F to 86° F by using a fan motor of 1 hp. Air-powered diaphragm pumps are used to move the jetted water and effluent through the process because of their ability to run dry and to handle large particles without any shearing action. A water supply/storage tank with a float valve is used

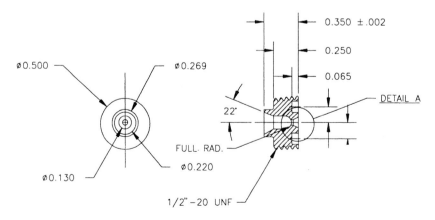

Figure 5.5. Pulsing nozzle inserts for rocket washout

Figure 5.6. Baffle tank used to recover propellant

to store the supply for the high-pressure pump while allowing makeup water to enter the system to replace any water lost in the process.

5.5 Operational Procedure

The system's operational procedure includes placing a rocket motor into the cradle for cleaning and the operator taking his or her position in a separate room with the control console as indicated in a complete layout of the system, as shown in Fig. 5.7.

Limit switches on the washout lance travel are preset for the size of work when setting up the operation. The control room has a closed circuit monitor to observe the operation from a safe enclosure. The high-pressure pump, air compressor, water storage tank, and water chiller are located between the control room and the removal equipment. A system start-up procedure is followed to bring the water recovery online and to start the high-pressure pump unit. Manually operated valves are opened to start the air-powered diaphragm pumps. From the console, the operator can control and observe the high-pressure pump, the lance extension and retraction, the nozzle rotation, the catch tank, and the equipment room. A cleaning cycle includes starting the system and operating a joystick to feed the rotating lance into the work and

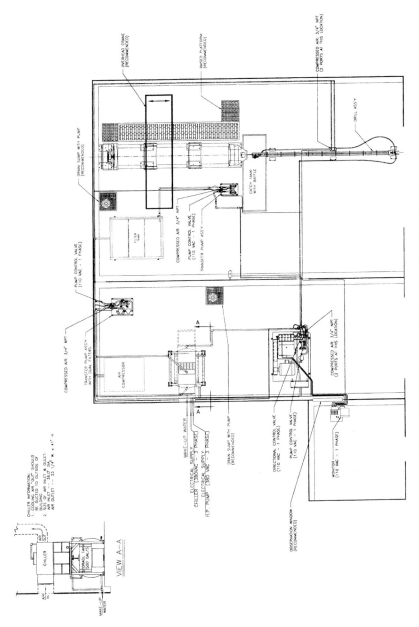

Figure 5.7. Rocket propellant washout system drawing

reversing the lance travel when the indicator light shows that a maximum travel has been achieved. The closed-circuit camera is located so that the process can be observed and controlled.

5.6 System Technology

The *Rocket Propellant Washout System* is a good example of existing water-jet and high-pressure pump technology being used to suit a customer's unique needs. Special engineering was necessary to design the cradle for holding the work and to allow for the integration of existing technology into the desired high-pressure system. Tubes were used to simulate rocket motors so that all system functions could be run and debugged, but rocket motors could not be provided to the manufacturing facility during the set-up testing period. The pressure and flow were tested on simulated material, so the removal of the propellant was ensured. The work done for previous systems, as described in Chapter 6, provided the background needed for developing a larger system like the *Rocket Washout*; reaching this level of technology could not have been possible without the previous work described in the references at the end of this chapter.

5.7 Mobile Equipment for Demilitarization

A company named Hydronautics in Laurel, MD, created equipment for removing high explosives (HE) from munitions, and in the 1980s, this equipment was shipped to Israel and Germany to be permanently installed near the storage sites of artillery shells and ammunition. Those involved in the development of the equipment had begun preliminary plans for a mobile unit to wash out munitions at various sites before Hydronautics closed its doors in 1988. Yeomans and Alba (1995) describe a multipurpose mobile plant for demilitarization to be provided by ALBA Industries in Munster, Germany, that is being designed to meet a growing need for demilitarization equipment.

5.7.1 Increasing Problem of Munitions Disposal

A general easing of tension between countries throughout the world has occurred, so there is a reduction in the need for some munitions. Obsolete weapons are being replaced by new technology, and new sales may include the purchase of the old munitions. Storage and maintenance are becoming more expensive, and there are environmental pressures to consider problems such as dumping munitions offshore, burying munitions, storing explosive shells, and recovering unexploded ordnance. The technology for reducing large quantities of the stored munitions does exist, and new methods can be developed to suit special situations.

5.7.2 Disposal Philosophy

ALBA Industries outlines considerations that must be approached for a good munitions disposal system:

- It is important to ensure that the disposal and/or destruction of the munitions are performed in an environmentally acceptable and economically viable manner.
- Munitions of varying types, sizes, and shapes should be handled.
- The technology, processes, and systems should be the best available.
- Relevant legislation, regulations, and local compliance should be understood and met.

5.7.3 Safety Aspects

Possible problems should be considered in the design and operation of a demilitarization system; consideration should be given to transportation of munitions, possible special handling of materials, personnel requirements, and possible secrecy requirements. In addition, the local population may express environmental concerns regarding the movement of munitions and the cleaning operation. Since munitions storage facilities are usually remote, some of the problems can be solved by bringing the equipment to the storage site instead of shipping the munitions to a cleaning facility. The mobile cleaning system is to be self-contained and easily maintained with the controls and personnel located at a safe distance from the cleaning operation. Procedures for handling equipment should consider the type of munitions being decommissioned, such as bombs, mines, rockets, ammunition, or detonators. Additionally, a key card system could be used to ensure that personnel such as fork truck operators are out of the area prior to the start of any system components. The cards can require that system personnel have left the work area and are at a safe distance from the operation.

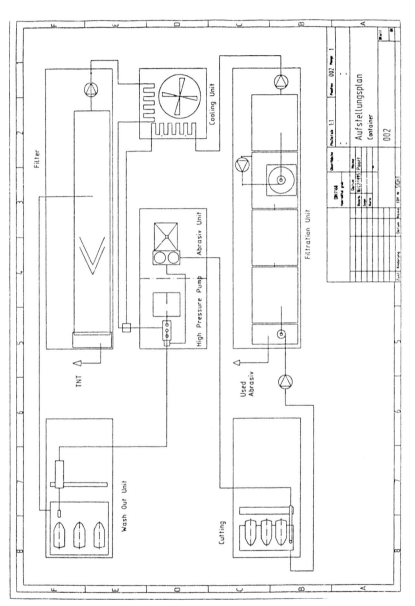

Figure 5.8. Rocket propellant washout diagram (*Courtesy ALBA*)

5.7.4 System Components

The system described by ALBA Industries uses a suspension jet abrasive unit capable of 300–1,000 bars with a jet size as large as 2.0 mm. The pump can be driven by electric motor or diesel engine as best suits portability. The cutting head is mounted on a single-axis tract or a three-axis manipulator as needed for the type of work. One method of cutting munitions involves a preloaded pallet of shells that can be cut open or defused by passing the head across several explosive warheads. A water/abrasive slurry catcher under the work receives the debris from the cutting operation as part of the containerized equipment concept. This stainless-steel catcher tank has baffles to protect it from the abrasive jet and has a cover to contain splash and overspray. A slurry pumping system moves the debris to a water-settling tank and then to a recycling unit. The water settlement and recycling system comprises a number of tanks arranged so that the water cascades from tank to tank to separate suspended particles. A hydrocyclone mounted over the initial tank is offered as an optional feature to help separate the solids from the water.

5.7.5 High-Pressure Washout

Certain types of munitions can be opened by unscrewing the fuse from the main explosive, while others must be cut open with the abrasive jet because of their age or design. Munitions that have been opened can be washed out with a nozzle lance that is centered in the shell, as indicated in Fig. 5.8. This diagram shows the washout unit, the cutting unit, and the filtration system. By containerizing a mobile plant for demilitarization, deteriorating stockpiles of unwanted munitions could be safely and efficiently processed near protected storage sites with a minimum impact on the environment.

References

Conn, A. F. (1986). An automated explosive removal system using cavitating water jets. *Twenty-Second DOD Explosives Safety Seminar*. Anaheim, California.

Gracey, M. T., and Conn, A. F. (1987). Cavitation erosion used for material testing. *Seventh International Conference on Erosion by Liquid and Solid Impact (ELSE VII)*. Cambridge, UK.

Gracey, M. T., and McMillion, B. (1995). *Rocket Propellant Washout System* using a pulsed nozzle. *Eighth American Water Jet Conference*, Houston, Texas.

Yeomans, M. J., and Alba, H. H. (1995). Multipurpose mobile plant for demilitarization. *Eighth American Water Jet Conference*, Houston, Texas.

Chapter 6

Munitions Decommissioning

6.0   Introduction

Using high-pressure pumps and water jets has proven to be a safe and economical method for removing high explosives (HE) from munitions and solid propellants from missile motors and rocket motors, as covered in Chapter 5. The company Hydronautics built complete systems for operations in Germany and Israel to wash out bombs and warheads in the late 1980s.

6.1   Decommissioning

During the filling of bombs and warheads, munitions can be incorrectly poured, which leaves cracks and voids in the explosive material. Although some of these HE materials can be melted, this process is slow and costly. Many nations are storing unwanted munitions that are either obsolete or too old to use because of the expense of melting out the HE and because of the restrictions on open burning or dumping them at sea. Demilitarization of munitions and removal of solid rocket propellant have led to the development of systems with water jets that have special characteristics to remove HE from 105- and 155-mm warheads. The heart of the system is a 150-hp (112-kW) high-pressure pump unit to produce 38 gpm (144 lpm) at 5,000 psi (34.5 MPa). Figure 6.1 shows the explosive washout facility layout, which includes the high-pressure pumps and major components of the system.

The components shown in Fig. 6.1 include:

1. High-pressure water pump unit
2. Water storage tank
3. Water chiller
4. Auxiliary pumps (a total of three)
5. Interconnecting piping, cables, and installation
6. Closed-circuit television camera and monitor
7. Settling tank with basket

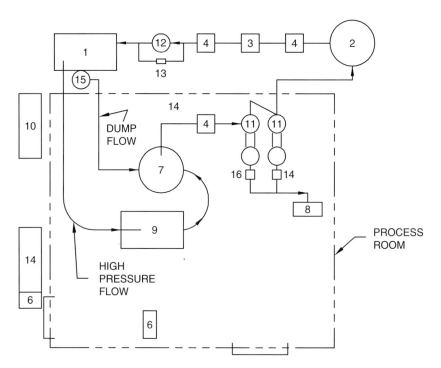

Figure 6.1. Munitions washout system developed by Hydronautics

8. Sludge container
9. Explosive washout unit
10. Hydraulic power unit
11. Centrifugal separators
12. Final filtration
13. Control console
14. Hoses, piping, and wiring
15. Auxiliary pump (only one)
16. High-pressure diverter valve

The components that come into direct contact with the HE are located in a process room, while the control console and power units are placed outside the room containing the munitions. After the warhead or rocket motor is placed in the washout unit, an operator runs the system and monitors the clean-out process from a safe distance at the control console. Each warhead is placed into the washout unit and unloaded from the process unit by hand, but this could be a robotic operation if desired.

6.2 Safety Features

The safety features of the demilitarization system inherently minimize the chances of either overstressing or heating the explosive materials. The system is operated by remote control, and the interlocks automatically shut down the operation if there is an error or a component failure. To maximize the safety of the washout system, these design features are incorporated:

- The high-pressure pump operates at about one-half the pressure required by conventional water-jet methods for HE removal.
- The water-jet velocities are well below the levels capable of detonating explosive material.
- The system provides a man-in-the-loop to observe the operation; this person can make decisions about any potential problems and has the power to shut down the system should a potential danger arise.
- The system is fully integrated with built-in safety interlocks for the lance travel and high-pressure water diversion to depressurize the system.
- Materials that come in contact with the effluent containing the explosive particles are stainless steel, brass, or plastic. The system

is fully grounded and uses plastic components, nylatron overlays, and other nonsparking materials in locations where impact sparking might occur.

• When the shell, motor, or bomb is being cleaned, the system maintains a fully flooded condition for the nozzle and explosive. The material removal is safer, more efficient, and foam free.

• Hydraulic and pneumatic actuators are used to move and rotate the system components, while the sensors and transducers are characterized by an intrinsically safe design.

To determine a safety factor for the operating pressure of a water jet to be used in an explosive washout system, the pressure needed to detonate a particular HE should be known. In a study conducted at the Royal Armament Research and Development Establishment (RARDE) at Fort Halstead, Sevenoaks, Kent, England, one detonation was reported during a series of tests with various RDX/Wax and RDX/TNT samples. It occurred when a sample of 91% RDX with 9% Wax Number 8 was impacted at a water-jet velocity of 1610 m/sec +/−10 m/sec. During 29 other tests at this velocity and 62 runs in the range of 1100 to 1600 m/sec, no detonations occurred. These extremely high velocities were achieved by a single-shot water cannon driven by a charge of propellant. The jet orifice was 12.7 mm in diameter, and the impacting slug of water was 115 ml. Each trial was filmed with a high-speed-framing camera. It seems that detonation pressures for sensitive explosives such as RDX, TNT, and Composition B are at least eight times higher than the nozzle pressures required to wash out HE; the safety factor may be as high as 40 or more.

6.3   Washout Unit

The explosive washout unit is shown in Fig. 6.2, and the principle parts are the water-jet cutting head, the lance and its drive, the shell rotation subsystem, and the effluent removal components.

The list of components for Fig. 6.2 include:

1. The shell (to be cleaned)
2. Water-jet cutting head
3. Water-jet lance
4. High-pressure water supply hose
5. Lance-bearing blocks
6. Lance-feed hydraulic cylinders

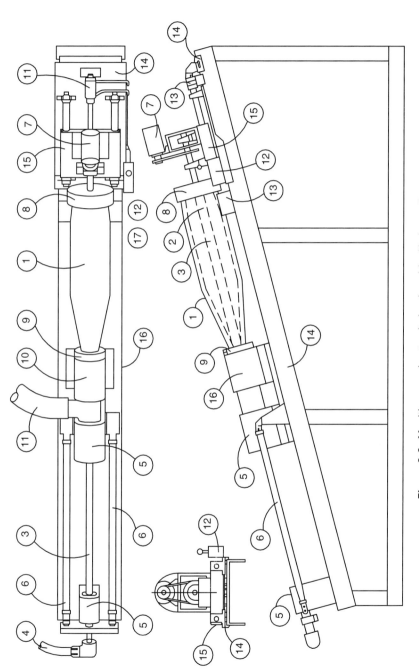

Figure 6.2. Munitions washout unit developed by Hydronautics

7. Shell-rotating hydraulic motor
8. Shell-base holder
9. Shell-nose holder (replaceable for each shell type)
10. Bearing block for shell spindle
11. Effluent-removal hose
12. Control valve for shell clamping
13. Shell-clamping hydraulic cylinder
14. Movable table (manually set for each shell type)
15. Sliding table for shell clamping
16. Main-bed plate
17. Guide (replaceable for each shell type)

The cutting head contains three self-resonating cavitating jet nozzles. The positions and orientation of these three nozzles were established in a development effort, discussed in a paper by Conn and Gracey (1988), which led to the present design. Two nozzles face forward to cut away successive segments of explosives, and the third nozzle is aimed backward to remove any remaining residues on the wall of the shell. The third nozzle cuts any chunks or material too large to exit the shell and provides a thrust balance so that there is no net lateral loading on the nozzle lance. The high-pressure water is supplied to the cutting head through a lance pipe, which is translated by two hydraulic cylinders. A pair of sealed, reed-type switches, which are repositioned for each munitions size to be cleaned, control the stroke length of these cylinders. All effluent water and explosive particles are fully contained during and after the washout cycle by the nose holder that is a molded soft plastic shell. During the cleaning operation, the explosive particles are flushed out of the shell through the hollow spindle-bearing block and into a large tee. A large hose from the tee allows the effluent to flow to the settling tank. Once the shell or motor has been placed into the washout unit and secured, further operations are performed from the control console.

6.4   System Controls

The control console is used to activate the system, for starting and stopping the shell rotation and for feeding or retracting the lance. From the console, the high-pressure pump and the hydraulic power unit can be started or stopped, and the pressure to the nozzle can be controlled. A diverter valve allows the pump to run while the water is recirculated until pressure is needed for the cleaning operation; the diverter valve is also part of the emergency interrupt feature when the functions are stopped by the operator. In case of a water pressure loss during the cleaning cycle, the system will automatically stop feeding the lance and shut down all control functions. This safety feature ensures that the cleaning head cannot be driven into the explosive when no water jetting occurs. Readouts on the control panel include the pump pressure, final filter pressure differential and status indicators for water flow, system pressure, and lance position.

6.5   Hydraulic Power Unit

The 3-hp hydraulic power unit provides the flow to the lance drive cylinders; shell rotation features of the unit include a Dennison pump and 20-g oil reservoir. The hydraulic valves are automatically actuated to feed and retract the lance during the cleaning operation.

6.6   Effluent Treatment

The effluent treatment components are part of a multistep approach for separating the particles of explosive material from the process water so that the water can be reused by the high-pressure pump. Figure 6.1, as already mentioned, shows the positioning of the effluent treatment components, which include the settling tank, air-powered pumps, centrifugal separators, water storage tank, and final filter. Each component removes successively smaller particles until only particles that are 5-μm or smaller remain in the water supply for the high-pressure pump. Under some conditions, a chiller is included in the system to remove excessive heat buildup created by high ambient temperatures and by the 150-hp high-pressure pump unit itself.

6.7   High-Pressure Pump Unit

The high-pressure pump unit, which supplies the cutting head, may either be driven by an electric motor or by a diesel engine. Typical pump specifications are 5,000 psi (34.5 MPa) at 38 gpm (144 lpm), requiring a 150-hp (112-kW) prime mover. The high-pressure pump should have a relief valve to protect from overpressuring, a system valve to allow running the pump under "no-load" until pressure is needed, a pressure read-out (e.g., a gauge), and automatic shutdown features. The automatic features can include the following:

- A switch gauge in the line to shut down the high-pressure pump in case of the loss of feedwater, which could damage the pump
- A system overpressure or underpressure switch gauge in case of a leak or nozzle wear that could affect the cleaning operation
- A switch gauge for the high-pressure pump crankcase oil so that low- or high-level monitoring can occur to protect the pump's power frame

Using the remote controls and built-in safety interlocks, the washout system can remove 16 lb of HE from a 155-mm warhead in well under 2 minutes, whereas a 105-mm warhead can be cleaned in 40 to 50 seconds. Figure 6.3 is a photograph of personnel reviewing the features of the explosive washout unit.

6.8   Abrasive Entrainment Systems

At the Eighth American Water Jet Conference in 1995, D. Miller of the BHR Group Limited explained the abrasive entrainment system called

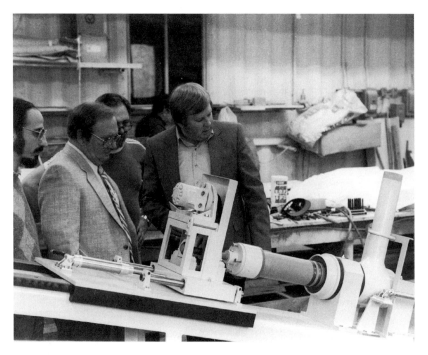

Figure 6.3. Photo of washout unit. L to R: Georges Chahine, Jim Terry, George Harrington, Mike Gracey (*G. Matusky*)

direct injection of abrasive jetting (DIAJET). DIAJET was developed as a cold-cutting tool for fuses, shells, bombs, mines, and rocket motors. Because the abrasive is mixed with water upstream of the cutting nozzle, the slurry can flow through standard hoses up to 3,000 ft (1,000 m). Acceleration of the abrasive in the nozzle is such an efficient process that standard water-blasting pumps can be used to drive the system.

6.8.1 European Versus U.S. Technology

Miller observed that abrasive water-jet cutting has been used more widely in Europe than in the United States for opening and cutting munitions. He explained that the ultra-high-pressure (UHP) water-jet abrasive systems were developed in the United States for manufacturing that requires high-quality installations. It has been more expensive and taken longer for U.S. organizations to build demilitarization facilities based on abrasive water-jet cutting because there is no need for precision cutting in opening munitions. Even though 80% of the abrasive cutting systems in the world are manufactured in the United States, there may be advantages to using direct abrasive injection over entrained abrasive systems for demilitarization.

6.8.2 Comparison of Entrainment and Direct Abrasive Injection

An entrainment jet is a turbulent mixture of air, water, and abrasive. A percentage of particles may have a radial velocity as they leave the mixing tube, which could make them strike the surface outside the area being cooled by the water jet. When cutting metals, the system could possibly cause sparking, which would not be desirable when cutting munitions. Entrained abrasive jets use UHPs of 3,000–4,000 bar produced by expensive intensifiers or pumps, whereas the direct abrasive injection systems use relatively low-cost water-blast pumping equipment. The entrainment systems usually require that the abrasive feed be located near the cutting head and that the direct abrasive head be located a great distance from the slurry generation module. Whereas the entrainment abrasive needs to be protected from moisture, the direct abrasive feed can operate in harsh conditions. The direct abrasive cutting system can have a cost advantage over entrainment systems and may be three to five times greater in effectiveness by using finer abrasives at lower pressures. Figure 6.4 shows a schematic of a DIAJET for cutting and explosive washout.

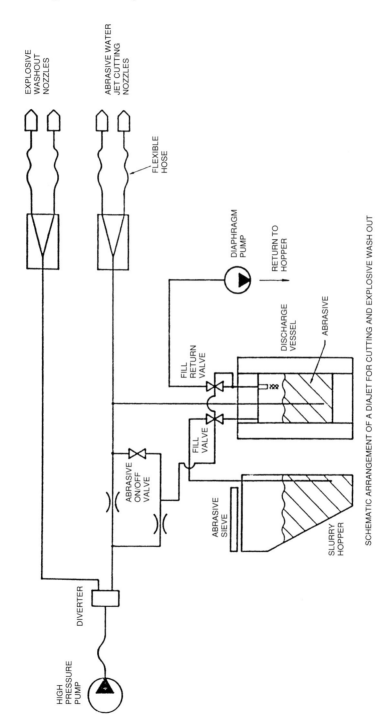

Figure 6.4. Diajet explosive cutting and washout diagram (*Courtesy of D. Miller*)

A pumping unit typically supplies 0.4–16 gpm (2–60 L of water per minute) at 10,000 psi (700 bar) to a slurry generation module with 10–20% of the water being diverted through a vessel containing abrasive particles. The water pressure causes the slurry to flow from the vessel to the cutting head that has a 0.016–0.12-in (0.4–3.0 mm) jet nozzle, depending on operating parameters. Miller suggested that there is a need to provide more high-pressure pump information to the demilitarization community so the technology can be better used.

6.9   Water-Jet Additives

6.9.1   Science Project

In 1996, the *WJTA Jet News* reported that a Minnesota high school student named Ana Navarro was doing research for a science and engineering fair in the area of the reaction of aluminum with water. Paul Miller of Alliant Techsystems explained to Ana's 10th-grade science class that there are hazards associated with aluminum powder when it mixes with water to form hydrogen. With equipment loaned by the Bureau of Mines and Ingersoll-Rand, Navarro tested the idea of mixing potassium phosphate with jetting water for a chemical passivation of the aluminum powder. Miller is currently using Navarro's idea in the process of cutting bombs and munitions. Knowledge of this phenomenon is vital for working with old aluminized explosives in the demilitarization industry.

6.9.2   Long-Chain Polymer

In the 1970s, Norman C. Franz of Flow Industries discovered the jet-stabilizing effect of additives in high-pressure water jets (Franz, 1976). The additives he used were of different type and chemical structure, such as glycerin, gelatin, and polyethylene oxide. The best effect was achieved with a linear long-chain polymer named *Polyox WSR-301*. Its use led to a tripling of the cutting efficiency and increased the jet-stream convergence.

The addition of a long-chain polymer can increase the effectiveness of water jets when cutting and removing HEs and propellants. A long-chain polymer is a linear, partially hydrolyzed polyacrylamide with a molecular weight of 14×10^6 to 18×10^6. The acrylamide and the acrylic acid function as dipoles—carbon atoms carry a partial positive charge, and oxygen atoms carry a partial negative charge. Water molecules are formed by negatively charged oxygen atoms that bond to the positively charged hydrogen atoms; in the case of polyacrylamide and polyacrylic acid, the water molecules are bonded to each other to form a longitudinal structure. A concentration of a 0.3% long-chain polymer can bond 13 to 14 molecules, and it is reported that its presence promotes laminar flow and decreases turbulence in water-jetting nozzles to form a more concentrated pattern. In a paper by Johnson *et al.* (2003), a long-chain polymer was used as part of a system to drill small inspection holes, a process that is called *Microintrusive Testing* of bridge foundations. The three main components described are a 5,000-psi (34 MPa) pump unit, a slurry supply accumulator, and a water-jet drill mechanism. A 2% long-chain polymer was used to hold the garnet in suspension for more than 1 month.

References

Conn, A. F., and Gracey, M. T. (1988). Using water jets to safely wash out explosives and propellants. *Ninth International Symposium on Jet Cutting Technology*, Sendai, Japan.

Franz, N. C. (1976). The interaction of fluid additives and standoff distance in fluid jet cutting. *Third International Symposium on Jet Cutting Technology*, Cranfield, England.

Royal Armament Research and Development Establishment (RARDE). Fort Halstead, Sevenoaks, Kent, England.

Fossey, R., Sims, K., Blaine, J., Tyler, J., Sabin, M., and Summers, D. (1997). Practical problems in the demilitarization of munitions. *Ninth American Water Jet Conference*, Dearborn, Michigan.

Johnson, P. W., Graettinger, A. J., and Sewell, C. H. (2003). An abrasive suspension water jet for drilling small diameter holes. *WTJA American Water Jet Conference*, Houston, Texas.

Miller, D. (1995). Abrasive water jets for demilitarization of explosive materials. *Eighth American Water Jet Conference*, Houston, Texas.

WaterJet Technology Association. (1997). Water jet additive makes demilitarization of munitions safer. *WJTA Jet News*, St. Louis, MO.

Chapter 7

Cutting Steel and Concrete

7.0   Introduction

Even though high-pressure water jets and cavitating nozzles can erode metal and many other materials, it can be advantageous because they allow users to increase the pressure and decrease the flow for cutting steel and hard materials. Abrasives are added to the ultra-high-pressure water stream to produce dimensionally stable cuts and promote the small kerf desired in hard materials, such as stone, steel, exotic metals, plastic, and concrete. When water is forced through a tiny opening and pressurized from 36,000 to 60,000 psi, it can cut soft materials, including food, paper, baby diapers, rubber, and foam. When abrasive particles are mixed into the high-pressure stream, hard materials, such as metal, composites, stone and glass, can also be cut.

7.1   Cutting Steel

Cutting steel with an ultra-high-pressure (UHP) water and abrasive jet was made popular with the advent of more dependable pumps and intensifiers. Water-blasting service companies developed special divisions to use the UHP water jets with abrasives to cut a variety of petrochemical industry materials, such as steel tanks, vessels, towers, concrete, alloys, and refractory. One of the special service divisions that uses UHP water-jet cutting performs large jobs, such as in the case of a petrochemical plant reactor in Houston, TX, as shown in Fig. 7.1. Before addressing large jobs, the equipment is developed and tested on smaller objects, such as on the tower shown in Fig. 7.2—the cutting head and track are attached with a cable and come-along.

Large jobs are either taken on a time and material basis or according to historical cost data. A paper by Zen and Kim discusses the viability of an abrasive water jet (AWJ) in cutting operations. Their work with the

Figure 7.1. Cutting steel and concrete job (*Courtesy of HydroChem*)

Figure 7.2. Testing steel cutting equipment (*Courtesy of HydroChem*)

machinability number and cutting rates could be used to estimate the cost of field jobs. Figure 7.3 shows equipment being installed on a 45-ft diameter reactor vessel with $\frac{7}{8}$-inch thick steel and 5 inches of refractory.

Figure 7.3. Installing cutting equipment on reactor vessel (*Courtesy of HydroChem*)

Figure 7.4. Cutting steel and refractory on reactor vessel (*Courtesy of HydroChem*)

The cutting shown in Fig. 7.3 took 18 hours to complete, and the cost to the customer was $21,000 (in 1995 dollars). Cutting of this type uses a 36,000-psi intensifier or a pump capable of up to 40,000 psi that would normally be diesel powered and trailer mounted for portability. Special jigs, fixtures, and devices have been developed by the industry to do linear cutting, circular cutting, and intricate designs. For the large jobs that include vessel cutting, equipment such as scaffolding and cranes as well as lifting equipment are required. Two-dimensional tables have been adapted to water-jet cutting heads to make designs and shapes in glass, rubber, wood, steel, leather, composite, aluminum, and plastic. In Fig. 7.4, a worker is monitoring the abrasive induction nozzle on the side of a vessel during the steel cutting operation. In Fig. 7.5, the top of the vessel is being lifted after the cutting operation is completed.

7.2   Cutting Concrete

Cutting concrete with UHP water has moved in many directions. In the late 1970s, the 36,000-psi intensifier pump was taken from laboratory conditions to the real and dirty world of contract cleaning work. It took four intensifier units available to the contractor to keep one working on the job. By the 1990s, the intensifier pump and its hydraulic

Figure 7.5. Lifting the top of a reactor vessel (*Courtesy of HydroChem*)

drive system had been improved enough to make the system a money-making tool in the contract cleaning business. Special divisions of certain service companies (as described in a previous example) take on concrete conduit cutting in industrial plants, cutting steel pipe, and cutting holes into concrete walls containing aggregate and steel rebar. High-pressure systems are commonly used for this type of work because less water is used; when abrasive is added, it will cut through almost any material with a narrow, clean kerf. In the 1980s and 1990s, the power end of the positive displacement pump was fitted with fluid ends that were developed to handle up to 40,000 psi. The UHP pump units were portable and less costly to maintain. Water-blasting service companies, however, still use 10,000–20,000 psi of water pressure for jobs such as cleaning towers, de-coking operations, heat exchanger tube cleaning (OD & ID), and the removal of polymers from process equipment (because lower pressures work well for this type of work).

7.3 Hydrodemolition

Hydrodemolition using high-pressure water blasters with rotating or reciprocating heads mounted on operator-controlled machines became popular in the 1980s to refurbish bridge surfaces and parking garage floors. Salt and chemicals used on the roads attacked the concrete and the steel-reinforcing bar used in the United States and other countries. Large systems were developed to carry the high-pressure pumps, water tanks, and robots used to accomplish the work. Some of the early attempts to cut through concrete entailed using water jets at pressures of 10,000 psi to gradually cut around the aggregate and reinforcing steel (rebar) with this method, a hole could be made in items such as storm drain junctions. Specialty nozzles, such as the *Cavijet*, would cut through the aggregate but still could not effectively cut through the steel rebar. Pressures of 10,000–20,000 psi were used to remove bad concrete on bridges and columns, but the pressures in the range of 30,000–40,000 psi were found to cut better than lower pressures. By adding abrasives to a UHP water jet, the jet could cut through the aggregate and rebar efficiently.

7.4 Pavement Cutting

Andrew Conn of Hydronautics, based in Laurel, MD, conducted tests that involved the cutting of pavement with cavitating water jets in 1985. It was part of a study supported by the Gas Research Institute (GRI) of Chicago, IL, to provide the gas industry with a multipurpose construction and maintenance vehicle (MCMV). Hydronautics took equipment to the streets of Baltimore, MD, to demonstrate that high-pressure water jets could cut city street pavement at a viable rate. The nozzle used for the field trial had a center body configuration, as shown in Fig. 7.6, that was designed to cavitate at 10,000 psi and 19 gpm, which was provided by an old Weatherford diesel-powered, trailer-mounted pump unit painted baby blue.

The aluminum nozzle containment box held a hydraulically reciprocating mechanism with a stroke travel that was easily adjusted for various cutting distances. After each run, the user had to remove the lead weights from the top of the box, move the unit and equipment to the next location, replace the weights, and then make the next cut. The cavitating nozzle cutting rate of 9 inches/min exceeded the 7 inches/min reported as the average rate for cutting with 35,000-psi abrasive jets on a reference pavement. The field demonstration appeared to have the potential of

Figure 7.6. Cavitating nozzle (*Stephen Gracey*)

reducing the cost of removing pavement segments for access to gas piping and utilities under street pavement, so a prototype of a commercially viable system was started in 1986. Additional equipment was designed and built that was used in further testing. This equipment included the circular pavement cutter, as shown in Fig. 7.7 (along with, left to right: Mike Gracey, Andy Conn, and George Arrington). Hydronautics was closed before the system was competed and put into service.

7.5 Machining Operations

Machining with water has become popular with machine shops of all sizes because of greater efficiency and productivity of implementing UHP water jets into some operations. To achieve the energy required for cutting materials, water is forced through an orifice (made from diamond, ruby, or sapphire) in diameter sizes of 0.004–0.022 inch. Hydraulically driven intensifier pumps that can handle pressures up to 87,000 psi and use PC-based software to translate drawings to the system are used for the cutting operation. Since its development in the early 1980s, the technology has grown rapidly as a machine tool. Water jets require few secondary operations and produce shaped parts with no heat-affected zones, heat distortions, or mechanical stresses that can be caused by other cutting methods. The jet cuts a narrow

Figure 7.7. Photo of circular pavement cutter (*G. Matusky*)

kerf with a satin-smooth edge and can provide better usage of material since parts can be tightly nested.

Abrasive is required for cutting hard materials, so a jetting system may include an abrasive hopper, gravity feed system, automatic on/off controls, abrasive mixing tube, and cutting nozzle. Abrasive travels from the hopper to a metering valve that supplies the abrasive through a precision disk to ensure consistent flow. When the cutting system is actuated, high-pressure water flows through the orifice into the cutting head and out through the mixing tube at up to three times the speed of sound. A venturi effect is created as the water enters the cutting head, which pulls the metered abrasive particles through the feed line and into the head. Abrasive combines with the water-jet stream to create the high-energy abrasive cutting stream exiting the mixing tube. The material being cut with the supersonic erosion typically produces thrust forces of less than 2 lb, so fixturing and special tooling are minimized. After the material is cut, a water-filled catcher tank collects the abrasive, cut material, and water while helping to abate noise. A catcher system can also be designed for clean and quiet underwater cutting for some products.

Flow International compares water jets with other cutting technologies by explaining how water jets expand a job shop's capabilities beyond what lasers, EDM, or plasma can provide. Water jets can cut virtually any material, whereas other cutting systems have their limitations. Water jets can cut reflective, conductive, or heat-sensitive materials and other materials, such as titanium, Inconel, brass, aluminum, glass, stone, composites, and steel. Lasers can cut a single layer of thin sheet metal with tolerances of ± 0.001 to ± 0.006 inch, but they must focus on what is being cut and are limited to the thickness of the material. Water jets can cut thicker materials with accuracy of ± 0.003 to ± 0.010 inch and are not dependent on the thermal properties of the material because the cutting is a cold, supersonic erosion process. Lasers have various parameters to adjust, such as those concerning optics and gases, while a PC-based controller automatically adjusts the only changing parameter for the water jet, which is cutting speed. Lasers also have the disadvantage of emitting toxic vapors when cutting some materials, while water jets produce no toxic by-products. When compared to wire EDM for cutting very thick materials, water jets cut faster at tolerances of ± 0.003 to ± 0.015 inch; EDM cuts with tolerances of ± 0.0015 to ± 0.0005 inch, but it cuts material using heat, so EDM cannot be used for every project.

Unlike lasers, plasma can cut through very thick material, and like EDM, plasma cuts faster than water jets. The accuracy for plasma is approximately 0.030 in for very thick material, but the cut is accomplished by using heat at temperatures of 20,000–50,000° F (10,000–27,000° C), so there is a significant heat-affected zone that may require secondary finishing. Water jets leave a good finish on the material with no heat-affected zone and no metallurgical changes. Water jets usually cut with a higher precision in the range of ± 0.003 to ± 0.010 inch; the range for plasma is a tolerance range of ± 0.005 to ± 0.030 inch.

Machine shops of all sizes are realizing greater efficiency and productivity by implementing UHP water jets into their operations; water jets are becoming the machine tool of choice for many shops. Since AWJ technology was first developed by Flow International Corporation (based in Kent, WA) in the early 1980s, the technology has rapidly evolved with continuous research and development. As a result, water jets are the fastest growing type of machine-tool cutting technology. In addition, water jets can cut virtually any material, leaving a satin-smooth edge, as mentioned previously. These benefits add up to significant cost savings per part in an industry that has traditionally defined productivity by cost per hour.

To achieve the energy required for cutting materials, water is pressurized and forced through a diamond, ruby, or sapphire orifice with a diameter ranging from 0.004–0.022 in. Hydraulically driven intensifier pumps pressurize water up to and sometimes more than 36,000 psi. Water is then carried to either an abrasive or straight water-jet cutting nozzle, which can be stationary or integrated into mobile equipment for cutting shapes or other intricate designs. PC-based software translates drawings to the system for the cutting head to follow a specific path.

7.6 Comparison of Water Jets with Other Cutting Technologies

As already described, the versatility of water jets expands a shop's capabilities beyond what is possible using only lasers, EDM, or plasma. While many cutting processes have their limitations, water jets can cut virtually any material, especially materials that are reflective, conductive, or heat sensitive. Water jets are ideal for cutting materials such as titanium, Inconel, brass, aluminum, glass, stone, composites, and any type of steel. Water jets provide a significant advantage when compared with lasers. Lasers can cut at faster speeds with tolerances of when cutting single layers of thin sheet metal. But, because lasers must focus on what is being cut, they are limited in the thickness of material they can cut. Water jets, however, can cut thicker materials. Because these jets cut with a cold supersonic erosion process, they are not dependent, as lasers are, on the material's thermal properties.

Wire EDM can cut very thick materials to more precise tolerances than can water jets. However, EDM cuts very slowly. Water jets can cut through very thick materials at a much greater speed than wire EDM. In addition, water jets are not dependent on the electrical conductivity of the material being cut as is wire EDM. Finally, just as in the case of lasers, wire EDM cuts with heat, so not every material can be cut using EDM.

Plasma, like lasers, cuts through materials faster than water jets. Unlike lasers, plasma can cut through very thick materials. However, since plasma cuts with extreme heat, there is a significant heat-affected zone on materials, so some secondary finishing is required in many cases. Water jets leave a perfect finish on the materials since they do not cut with heat, so there are no heat-affected zones and no metallurgical changes. Many job shops are adding water jets to their operations as a complement to other cutting technologies such as EDM, laser, and plasma systems. In fact, some are even replacing traditional cutting methods with water jets. While each shop has its own requirements for cutting projects, most are finding water jets to be a tremendous benefit to their operations—enhancing both productivity and profitability.

7.7 Future Technology

A significant new development in abrasive water jetting is Flow International's technology release of the *Dynamic Water Jet*. This jet helps manufacturers cut parts up to four times faster than traditional water jet cutting, and it also eliminates taper to produce higher quality parts. By using sophisticated computer models to control a small, articulated "wrist" cutting head, *Dynamic Water Jet* automatically adjusts cutting angles to cut parts at high accuracy and speed. Potential markets of interest for this technology focus on existing users of flat-stock cutting technology or organizations with applications that were previously inaccurate, too slow, or too expensive to cut using abrasive water jets. In the *Quality Waterjet Newsletter* (www.qualjet.com), a discussion of water jet terminology points out that people often incorrectly believe the word *water jet* refers to an *abrasive jet*. Multiple spellings also exist for the term *water jet*, such as *waterjet* or *water-jet*. The Website www.waterjets.org offers information to novices and experts in the abrasive-jet and water-jet machining business.

References

Conn, A. F. (1986). Rapid cutting of pavement with cavitating water jets. *Eighth International Symposium on Jet Cutting Technology*, Durham, England.

Gracey, M. T., and Smith, R. E. (1995). Cutting steel and concrete with ultra-high-pressure water and abrasives. *Eighth American Water Jet Conference*, Houston, Texas.

Matthew, M. (2005). *Quality Water Jet Newsletter*. Bellevue, WA:

Miller, D., and Claffey, E. (1995). Improving the quality and the speed of abrasive water jet cutting. *Eighth American Water Jet Conference*, Houston, Texas.

Chapter 8

Stripping and Surface Preparation

8.0   Introduction

High-pressure pumps using various nozzle configurations have been used to clean surfaces since the 1950s, as discussed in Chapter 4. Nozzles that allowed the injection of sand and other abrasives were in use by the late 1960s to 1970s. These nozzles were often used on the end of a handheld water-blasting gun to remove coatings, rust, and scale from metal or to recondition structures, such as public buildings, statues, and masonry. Water alone is good for removing material from a hard substrate, such as hard paint over steel, but water also tends to cut into a soft surface, such as hard paint over concrete or the grout between brick structures. The addition of abrasive in the water jet can erode a hard coating from a surface without cutting into the soft substrate.

8.1 Wet Abrasive Blasting

Since the 1970s, the field of high-pressure pumps has made an effort to replace dry sandblasting with high-pressure water sandblasting. The following list presents a procedure for comparing the two methods. Tests were conducted using a high-pressure pump with a pressure of 7,000 psi at 12.5 gpm with sand eductor and then using a dry sandblasting rig with a $\frac{3}{8}$-inch nozzle and 110-psi air pressure:

- The cleaning rate for wet abrasive blasting was approximately 150 ft^2 per hour for very badly rusted surfaces, such as ship hulls and tank bottoms with large rusted areas, and for surfaces with tars and heavy paint. On very light rust and mill scale, such as vertical steel surfaces and bridges, the rate was approximately 150 ft^2 per hour.
- The cleaning rate for dry sandblasting was approximately 135 to 140 ft^2 per hour for very badly rusted surfaces, such as ship hulls and tank bottoms with large rusted areas, and for surfaces with tars and heavy paint. On light rust and old paint, such as on the sides of tanks and steel buildings, the cleaning rate was approximately 160 ft^2 per hour.

In actual practice, the cost of both methods was about the same after considering the price of sand, labor, scaffolding, and cleanup. The wet abrasive blasting had the advantage of not emitting any sand dust, which was a health consideration of dry sandblasting. In spite of all the research, demonstrations, and efforts by manufacturers and contractors in support of wet abrasive blasting, the technique failed to be readily accepted throughout the 1980s as a surface preparation method to replace dry sandblasting. Several companies began to offer a new water-blasting media system to use with conventional high-pressure water-blasting pumps for removing stains, light buildup, discoloration, and graffiti, for which sandblasting was too aggressive. The system used a nontoxic, nonhazardous, biodegradable, water-soluble media, which was an industrial grade baking soda. By the 1990s, this biodegradable media used for blasting was a big hit in the refineries and petrochemical plants for cleaning concrete, aluminum, insulation, piping, vessels, and other such items. Table I shows the approximate media usage for various pump flows and pressures.

Increasing the amount of media did not seem to increase productivity, while standoff distances varied depending on the material being

Table I
Industrial-Strength Baking Soda Media for Use with High-Pressure Pumps

Pounds per Minute	Gallons per Minute	Pounds per Square Inch
0.5 to 0.75	0.15 to 2.0	3,000 to 15,000
1.9 to 1.25	4.0 to 6.0	1,500 to 5,000
1.5 to 1.75	8.0 to 12.0	5,000 to 12,000
2.0 to 2.5	14.0 to 16.0	5,000 to 12,000

removed. Pressurized paint spray cans have been around since the late 1950s, but their use by gangs for graffiti began to increase during the latter part of the twentieth century. Graffiti artists climb structures and equipment to post their names, organization's identification, or message, which in turn expanded the business of graffiti removal. Various chemicals, baking soda, abrasives, glass beads, and dry ice have been used with high-pressure pumps to remove the work of the nighttime vandals. A series of articles discussing cosmetic cleaning and graffiti removal with various media have appeared in *Cleaner Times* and other publications.

8.2 Chemicals for Use with High-Pressure Pumps

Chemicals used with high-pressure water-blasting pumps are part of a broad and ever-changing discussion. Chemicals that were first used years ago have been discontinued for ecological reasons. Chemicals that meet EPA requirements have been introduced to the industry; these modern chemicals include various types of soap, acids, baking soda, and foams. In most cases, a high-quality, concentrated detergent will suffice, and hot water at a temperature of 140° F or higher can double the effectiveness of cleaning chemicals. Cold water can cause oil and grease to harden, so it is difficult to remove substances in the range of 1,000–4,000 psi. To avoid using heat and chemicals in the heavy industrial market, water-blasting contractors use pressures in the range of 6,000–10,000 psi to remove materials from tanks, vessels, piping, and heat exchangers. Much of this type of work does not require chemicals, additives, or abrasives. Some projects, however, could benefit from using additives. For example, when carbon steel is exposed to the atmosphere by ultra-high-pressure (UHP) pumps using water pressure in the range of 30,000–40,000 psi, a rust inhibitor might be desirable to fight flash rust. A sophisticated paint coating system may require a white metal or a near-white metal blast with a designated anchor pattern for adhesion.

Acid inhibitors were used as early as 1946 to prevent rust accumulation on water-blasted surfaces and to reduce the hydrochloric acid reaction on steel. Monosodium and disodium phosphates plus metasilicate were used to passivate and seal freshly cleaned steel. By 1951, bare steel was sprayed with a dilute phosphoric acid solution; later, a 0.2% mono/disodium phosphate and sodium silicate solution was used when cleaning and removing rust from steel. It was possible to add this solution to the suction tank of the water blaster, or it could be sprayed later during the derusting process. Many other types of chemicals are available for derusting and holding steel when pressurized water is used to remove coatings and when steel is prepared for recoating.

8.2.1 Hold Tight Solutions

Peter Petkas of Hold Tight Solutions explained that a good product can virtually eliminate detectable chlorides and other salts from the blasted surface and will eliminate even the light blush or light flash rust that often appears after UHP water blasting. For a flash-free window during the 48–120 hours after the blast cycle, the proper water dilution of the product can be used to hold the surface for coating. Some inhibitor products can be used in the water supply to the UHP pump if the manufacturer approves. Otherwise, a washdown cycle can be used to apply the solution if a surface is highly contaminated with chlorides or if the water is very hard. In addition, if the surface is deeply pitted or the weather conditions are marginal, the solution concentration can be increased for the desired effect.

8.2.2 Polymeric Additives

In 1974, a company was formed that started promoting the use of a long-chain polymer to improve high-pressure water jets. First developed for cleaning heat exchangers in a West Coast refinery, the long-chain polymer has been used in applications such as cutting, drilling, lancing decontamination, fracturing, descaling, and surface preparation. The manufacturer contends that the product reduces costs of operating and equipment maintenance, reduces the horsepower required, and extends nozzle life. Louis *et al.* (2003) explained that the jet stability significantly improved with the use of polymers, and polymers have been successfully applied to abrasive water injection jets. Water alone was replaced by a long-chain polymer (a partially hydrolyzed polyacrylamide) solution. Cutting efficiency could be increased by 20%, and abrasive consumption could be reduced from 25 to 30% by using a cutting system optimized for the use of polymer solutions.

8.3   UHP Water Jetting

Ultra-high-pressure water-jetting (UHPWJ) equipment combined with intensifiers and pumps began to appear in the 1970s and became increasingly dependable with use. Water- blasting contractors began to use the trailer-mounted, diesel-driven hydraulic over water intensifier units to approach jobs that could not be performed or were difficult to address at lower pressures. The early field equipment was impossible to keep running, but more dependable intensifier units and positive displacement pumps were created in the 1980s. The 36,000-psi units were also used to increase revenue in a sagging economy; by the 1990s, however, the dependable pump units had inspired an array of useful UHP accessories. Rotating handheld guns with sapphire nozzles, tube lancing equipment, and automated surface-cleaning mechanisms were developed for use with positive displacement pumps that could deliver up to 40,000 psi of water pressure.

8.4   Case Study

Using high water pressure is not always the correct answer to cleaning and stripping even though thousands of jobs have been performed using UHPWJ equipment. Every job requires a different amount of pressure, as the following example suggests. Figure 8.1 shows an offshore rig docked in a shipyard near Tampico, Mexico, in 1998. Coatings needed to be removed from its surface. Gracey (1999) describes and discusses the positive displacement pump, flow splitters, tumble boxes, and rotating guns used in this coating removal job.

Because of articles and advertising that indicated a need for 36,000 to 40,000 psi at 3 to 4 gpm, the client who requested that the coating on the rig be removed wanted to purchase equipment that would operate at 40,000 psi and 8 gpm. The dual gun operation specified by the client required the use of 2 flow splitters, 2 tumble boxes, 4 air hoses, 1 interconnecting hose, 18 sections of 40,000-psi hose, 2 rotating handheld

Figure 8.1. Coating removal job on Russian semi-submersible in Mexico

guns, and spare parts (just to get started). An operator being trained to use the rotating guns is shown in Fig. 8.2, two air-operated rotating barrel guns after testing are pictured in Fig. 8.3, and the 40,000-psi

Figure 8.2. Operator testing ultra-high pressure handgun for coating removal

Figure 8.3. Dual ultra-high pressure handguns

diesel-powered pump unit to be used for coating removal after shop testing is depicted in Fig. 8.4.

Only 1 week after the start-up and training began, the equipment started breaking down, and a technician had to return to the job site to

Figure 8.4. 40,000-psi pump unit for coating removal job

Figure 8.5. 40,000-psi hose being inspected for damage

repair the pump and gun swivels while reviewing the high-pressure hose problems. The rotating guns soon began to break shafts and use seals at an unsatisfactory rate, while sapphire nozzles were being used by the dozens. Figure 8.5 shows the 40,000-psi hose on the job site that was eventually replaced because of ruptures.

After 1 week of operation, the first fluid-end stuffing box cracked, and the boxes continued to crack about every 40 hours; the work could not proceed with the 40,000-psi equipment. A 30,000-psi pump unit was sent to the job site and proved to be effective for removing the epoxy coating system. The accessories that were rated to work well at 40,000-psi working pressure actually performed better at 30,000 psi, so the client was eventually satisfied with the reduced operating pressure. The 40,000-psi systems have continued to improve since 1998, but the latter example suggests that the minimum pressure necessary to do the work should be used. Figure 8.6 shows the coating that was removed from the offshore rig by the high-pressure water jets at both 30,000 and 40,000 psi. Using these two water pressures for the same job revealed that operating at 40,000 psi cost more than operating at 30,000 psi and that multiple guns can increase operating costs.

Figure 8.6. Coating removal on semi-submersible

8.5   Increasing Pressure

As UHP pumps and accessories develop, the trend is to increase operating pressures to more than 40,000 psi, with the possibility of 55,000-psi being the next plateau. A report by Jet Edge compares surface preparation rates at 40,000 and 55,000 psi for a water storage tank in Essington, PA. Table II shows the test results when using handheld air rotating guns at the two pressures but different flows.

Table II
Surface Preparation Rates for Different Operating Pressures

Water pressure (pounds per square inch)	40,000	55,000
Flow rate (gallons per minute)	6.0	4.3
Reaction force (pounds)	62.4	52.4
Stream velocity (feet per second)	2436.0	2856.0
Average square feet per hour	87.6	178.0
Gallons per hour	360.0	252.0
Gallons per square foot	4.1	1.4
Gallons per 8-h day	2880.0	2064.0

The results in Table II were attained using a Jet Edge 55,000-psi *36-250DX* unit with identical manifold, the same operator, and a hand-held air-rotating gun. The primer coating was *60P1* lead-based primer (1–2 mm thick), the top coating was single-component vinyl (three coats from 4–5 mm thick), and the substrate was a steel water storage tank in Essington, PA. Table III shows the test results when using a mechanical crawler to remove the lead-based primer and vinyl coating system. The tests were performed using a 55,000-psi Jet Edge *36-250DX* unit with 330 hp and 12 gpm. A Jet Edge *Hydro-Crawler* with identical spray bars that have orifices for desired flow was used.

Table III
Surface Preparation Rates with a Mechanical Crawler and Different Pressures

Water pressure (pounds per square inch)	40,000	40,000	55,000
Flow rate (gallons per minute)	12.0	6.0	4.3
Reaction force (pounds)	124.8	62.4	52.4
Stream velocity (feet per second)	2436.0	2436.0	2856.5
Average square feet per hour	425.0	199.0	520.0
Gallons per hour	720.0	360.0	252.0
Gallons per square foot	1.7	1.8	0.5
Gallons per 8-h day	5760.0	2280.0	2064.0

8.6 Improving Water Jet Standards for Surface Preparation

Frenzel (1997) discusses the standards for surface preparation of concrete, steel, and coating removal. The standards of third-party organizations, such as NSRP, SSPC, NACE, ASTM, and ISO, should provide a common language to define the process. Frenzel states that it is vital for the coating industry to understand what water jet pumps have to offer; in turn, it is also important for the water-jetting industry to understand the demands of the coating industry concerning substrate characteristics.

8.6.1   Water-Blasting Equipment

Water-blasting equipment has been used since the 1960s to clean loose residues, chalking, and dirt from surfaces; the water-blasting method does not produce dust like the conventional dry-blasting method for accomplishing the same tasks. Because of environmental concerns and the evolution of new equipment, which is cost-effective and reliable, using high-pressure water jets is more popular than in the earlier years. Water by itself or with soluble abrasives is used for maintenance removal of heavy rust, old coatings, rubber, salts, and invisible contaminants that may not be present in new construction work. Using water jets and hard abrasives is a common practice in new construction where builders are generally concerned about the dust levels. The NACE/SSPC surface preparation standard includes a statement that water jetting is primarily used for recoating or relining projects because it does not provide the primary anchor pattern known to the coating industry.

8.6.2   High-Pressure Water Jetting

High-pressure water jetting (HPWJ) is performed with high-pressure pumps at pressures from 10,000 to 30,000 psi. The velocity increases to the speed of sound and starts to change the amount of cleaning that can be accomplished by water alone. Some contractors use up to 20,000 psi to remove lead-based paint from tanks in plant facilities where the effluent can be processed by the plant's treatment system.

8.6.3   Ultra-High-Pressure Water Jetting (UHPWJ)

UHPWJ refers to cleaning that is performed at pressures above 30,000 psi, as previously mentioned. UHPWJ equipment generally uses less water at ultra-high pressures to remove tightly adhered coatings that HPWJ does not remove. HPWJ, using a larger mass of water, removes less tightly adhered coatings and rust, so a contractor may use both HPWJ and UHPWJ to approach jobs with different requirements.

8.7   UHPWJ for Coating Removal

Major coating manufacturers have realized the advantages of using UHPWJ for certain surface preparation jobs (Schmid, 1997). UHPWJ is sometimes the preferred method for tackling surface preparation because of new coatings developed for use with the technique; in addition, UHPWJ allows for better coating adhesion and lower levels of residual chlorides and sulfates. Modern projects that use UHPWJ for surface preparation include coating removal on bridges, storage tanks, ships, and large complex steel structures.

8.7.1   Removal Rate Improvement

Productivity of older handheld water jet lances were once one quarter to one third that of handheld dry abrasive nozzles. In the past, UHPWJ coating removal rates were generally 20–100 ft^2/hour with a handheld tool, whereas productivity for a dry abrasive nozzle was about 90–120 ft^2/hour. Slow nozzle rotational speeds and operating pressures in the range of 30,000–35,000 psi were said to limit the removal and cleaning rates for UHPWJ. Technology rapidly advanced between 1992 and 1996 to meet the demand by increasing pressures to 40,000 psi and increasing nozzle speeds to the range of 3,000–3,500 rpm. Improvements in nozzle technology also helped to raise productivity rates and make them comparable to rates for handheld abrasive blasting; typical rates of 80–200 ft^2/hour could now be compared with 90–120 ft^2/hour for abrasive blasting on similar materials.

8.7.2   Equipment Improvement

Harsh conditions in the field for surface preparation equipment made it difficult for original UHPWJ equipment to withstand the demands of shipyards and work on bridges and storage tanks. Attempts to convert hydraulic intensifier pumps from factory applications to unpredictable conditions were marginally successful. The early portable intensifiers were unreliable and complex, requiring trained technicians to maintain the equipment. Equipment advances and improved reliability have contributed to the rapid acceptance of UHPWJ in the high-pressure pump field. The evolution of positive displacement plunger pumps has been the single most significant development for the industry. For many years, positive displacement pumps have been used for water blasting, using pressures in range of 10,000–20,000 psi. These pumps are generally easier to maintain and easier to operate in dirty field conditions than intensifiers, so they have been a popular choice for projects in harsh industrial environments. The development of UHP versions of the positive displacement pump (crank-driven pumps) now reliably operate and use water pressures up to 40,000 psi in field conditions. A high cost of the equipment and hourly operation is a reality, but at least field personnel can effectively use UHPWJ for contract cleaning and surface preparation.

References

Frenzel, L. M. (1997). Continuing improvement initiations of surface preparation with water jetting. *Ninth American Water Jet Conference*, Dearborn, Michigan.

Gracey, M. T. (1999). Using 40,000 water jetting for field work. *Tenth American Water Jet Conference*, Houston, Texas.

Louis, H., Pude, F., and von Rad, C. (2003). Potential of polymeric additives for the cutting efficiency of abrasive water jets. *WJTA American Waterjet Conference*, Houston, Texas.

Schmid, R. (1997). UHP water jetting gains acceptance for surface preparation. *Ninth American Water Jet Conference*, Dearborn, Michigan.

Chapter 9

Environmental and Safety Concerns and Improvements

9.0 Introduction

Environmental and safety concerns regarding high-pressure pumps include any pollutants caused by the pump or system and protection for personnel who work in the proximity/vicinity of the pump when it is being operated. The fluid being pumped may be hazardous, so equipment and methods have been developed to monitor and to control accidental release of products into the atmosphere. In addition, the field has taken steps to improve training and ensure personal protection; these initiatives promote safety for those working with high-pressure pumps.

9.1 The Environment

The environment was not of prime concern in the United States until air and water pollution reached an alarming level. By the 1960s, places such as Boston Harbor were like sewers, and waterways like the Houston ship channel were fire hazards. A 1970 issue of the *Lamar Engineer* (published by the engineering department of Lamar University in Beaumont, Texas) featured cartoons and editorials that reflected the conditions of the air, water, and natural resources. Figure 9.1 shows a somber cartoon similar to those used in the 1970 issue to spur people's thoughts about how severe conditions were at that time.

The Water Pollution Act of 1948 was totally revised in 1972 with the goal of eliminating all pollutant discharges into United States waterways by 1985 so that they could be fishable and swimmable. Those who

THE NATURE OF MAN

Figure 9.1. *Courtesy of the Lamar Engineer*

fished, swam, ate, and breathed in the United States could soon enjoy the improvements. An article in the April 2005 *Reader's Digest* compared the conditions in the 1970s to those in the early twenty-first century and stated that the United States' air is cleaner, its lakes are purer, its forests are healthier, its endangered species are recovering, and its toxic emissions have been reduced; in addition, acid rain has diminished dramatically. Nearly all environmental trends in the United States are positive and have been for years. Boston Harbor is sparkling again, and the Potomac River, which gave off a stench in the 1960s, is lined with thriving waterfront restaurants. The Chicago River was a virtual open sewer in the 1960s and now hosts charming dinner cruises. The credit for improvements in protecting the environment is given to ecolegislation, green organizations, corporate cooperation, and new inventions.

9.2 Hazardous Waste

Hazardous waste may be in the form of liquids, solids, or sludge, so the waste includes lubricating oils and products being pumped by the systems being discussed in this chapter. The U.S. Environmental Protection Agency (EPA) hosts a HAZTRACKS Web page (*http://www.epa.gov/earth1r6/6en/h/haztraks/haztraks.htm*) that provides information on the U.S.-Mexico Hazardous Waste Tracking System. Hazardous waste can be a by-product of manufacturing processes or commercial products, such as cleaning fluids or battery acid that have been discarded incorrectly. A hazardous material exhibits one or more of the following characteristics:

- **Ignitability D001:** Ignitable wastes can create fires under certain conditions. Examples include liquids such as solvents that readily catch fire and friction-sensitive substances.
- **Corrosiveness D002:** Corrosive wastes include those that are acidic and those that are capable of corroding metal, such as metal tanks, containers, drums, and barrels.
- **Reactivity D003:** Reactive wastes are unstable under normal conditions. They can create explosions, toxic fumes, gases, or vapors when mixed with water.
- **Toxicity D004-D042:** Toxic wastes are harmful or fatal when ingested or absorbed. When toxic wastes are disposed of on land, contaminated liquid may drain or leach from the waste and then pollute ground water. Toxicity is identified through a laboratory procedure using the Toxicity Characteristic Leaching Procedure (TCLP) test.

9.3   Environmental Protection

Petrochemical industry water-blasting methods that support environmental protection have improved in the late twentieth and early twenty-first centuries for several reasons. In the past, a vehicle could be washed or a water-blast job could be done in an industrial plant with little concern about the dirt/debris generated, but the EPA now dictates that no contaminated water should leave the premises on which the job is performed. In the past, petrochemical plants usually received the effluent from work done by contractors cleaning their heat exchangers, towers, and process equipment, but now effluent has to be dealt with properly. Forms must be filled out and procedures must be followed for local agencies. Additional costs exist for processing the used water, for hauling the waste, and for disposing of the waste properly.

In the past, the plant or client provided large quantities of water needed for the jobs, but because of the cost of processing the water, contractors were encouraged to develop equipment to recirculate the effluent and reuse the water for their cleaning operations. If a water-blast pump produces only 20 gpm, a 10-h operation could use as much as 12,000 g of water. A recirculating filtration unit can process the effluent for reuse while concentrating nonsoluble debris for more economical disposal. This equipment was developed for the petrochemical process equipment, such as towers and heat exchangers, but can be adapted to projects involving ship cleaning, rail car cleaning, and tank-truck washing.

Another problem for the plant and the contractor involved working with dangerously contaminated items like some heat exchangers. The product on one particular heat exchange to be cleaned was known to cause cancer, heart and respiratory problems, or liver damage. As a result, personal protective clothing was developed called "acid suits" that enabled personnel to work on hot tube bundles and tanks without being exposed to the toxins. The operation, however, was expensive and uncomfortable for the worker, and sometimes the clothes had to be disposed of when the job was completed. This type of work prompted the development of portable containment systems to control the liquid effluent and the gases produced in the cleaning operation. A tube bundle can be placed inside a roll-on/roll-off enclosure that contains the tube lancing and shell side-cleaning hardware. Figure 9.2 shows a hazardous material containment system used in high-pressure water-jet cleaning of heat exchangers.

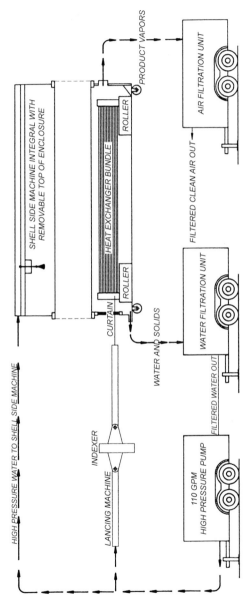

Figure 9.2. Hazardous material containment system developed by Hydro-Services

Tower cleaning and cleaning of process equipment, such as vertical heat exchangers, are types of jobs where the effluent can be captured. Water consumption can be reduced by as much as 1800 gph, and disposal is greatly simplified. The solids are separated from the water, and personnel's exposure to the effluent is less than the exposure associated with previous methods of cleaning. Figure 9.3 shows effluent reuse in tower cleaning. Figure 9.4 shows a process equipment cleaning arrangement used for vertical heat exchangers.

9.4 Pump Pollution

Pollution caused by pumps has also come under scrutiny in the last few years. A leak in the high-pressure packing causes environmental pollution, safety hazards, loss of energy, and costly clean up. For example, sucker rod pumps can leak oil on the ground and into waterways, as shown in Fig. 9.5. The patented technology called the *Hydro-Balanced Packing System* (H-B), as shown in Fig. 9.6, was developed by Harold Palmour to eliminate the pollution caused by pumps in the oil patch.

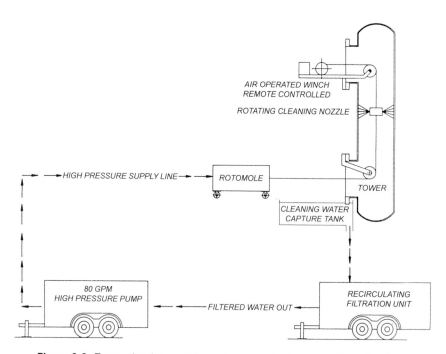

Figure 9.3. Tower cleaning containment system developed by Hydro-Services

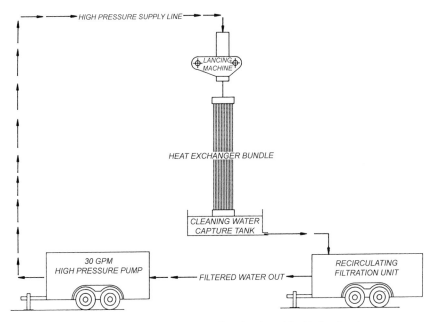

Figure 9.4. Vertical lancing operation developed by Hydro-Services

Figure 9.5. Fluid Hydro-Balanced rod pump seal (*H. Palmour*)

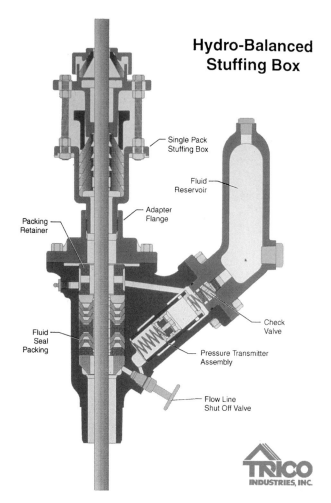

Figure 9.6. Grease Hydro-Balanced rod pump seal (*H. Palmour*)

The technology for the H-B system transfers the pressure of the product being pumped to a sacrificial barrier fluid of known characteristics. When the seal wears sufficiently to leak, only the environmentally friendly fluid leaks to the atmosphere. The technology can be built into new equipment or can be retrofitted to pumps in service with the proper modifications and sealing assemblies. During the development of the H-B system for high pressure pumps, a test loop was used to do a proof of principle to 10,000 psi as shown in Figure 9.7 using a horizontal, triplex, plunger pump. Figure 9.8 shows a stuffing box from the test

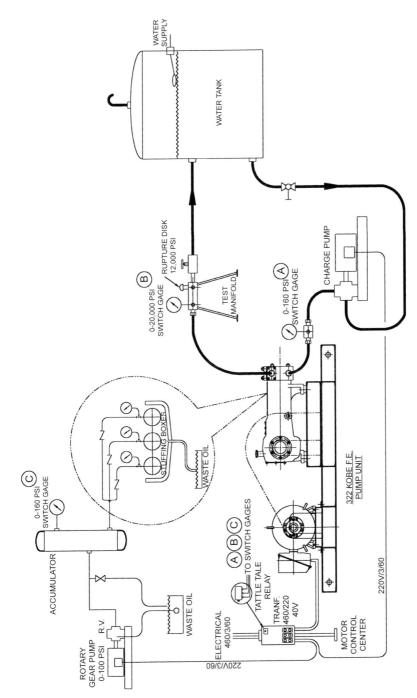

Figure 9.7. Hydro-Balanced packing system test loop diagram

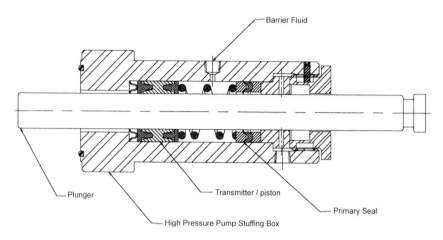

Figure 9.8. Hydro-Balanced packing arrangement for test loop

pump with the transmitter/piston that transfers the pump pressure to the barrier fluid. The other benefits of using the H-B system can be better packing life, control of the plunger lubrication, and a decrease in environmental pollution. The H-B packing arrangement was installed in a National pump in 2000. The test showed promise for pumping non-lubricant fluids.

9.5   Pollution Control

When ordering high-pressure pumps in offshore applications, oil companies have started asking for pollution control features on the equipment. The offshore-type skid with drip rim and drain pan have been used for many years, but now the high-pressure pump has been outfitted with double-packing systems, oil-catch mechanisms that signal when the catch vessel is full, and instruments that determine if the packing is leaking. Around the year 2000, five pump units were built for a large oil company that incorporated a system that collected the plunger lubricating oil and included a pump shutdown if the packing experience was a catastrophic failure, as shown in Fig. 9.9.

The weight of the liquid collected in the catch vessel eventually would engage a microswitch to initiate the alarm and shutdown. The liquid could be drained and/or the packing repaired for the continuation of the pumping operation. By 2003, other oil companies insisted on having *Smart Hart* instrumentation with the pumps that they ordered for offshore operation, including methanol pumping. Figure 9.10 illustrates a system with a *Smart Hart* instrument that reads the pressure

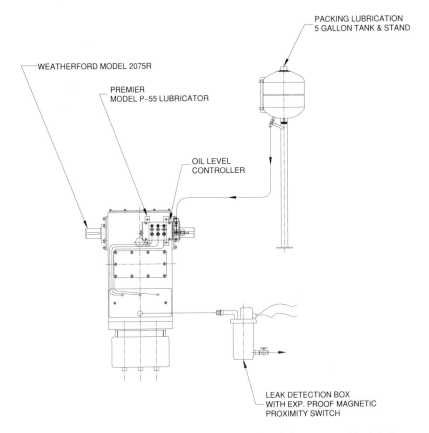

PACKING LUBRICATION
5 GALLON TANK & STAND

WEATHERFORD MODEL 2075R

PREMIER
MODEL P–55 LUBRICATOR

OIL LEVEL
CONTROLLER

LEAK DETECTION BOX
WITH EXP. PROOF MAGNETIC
PROXIMITY SWITCH

Figure 9.9. Pollution prevention control system for triplex pump pot installation

between the primary and secondary seals in each stuffing box of a triplex plunger pump. The sealing arrangement used with the *Smart Hart* instrumentation is a double-packing configuration, as shown in Fig. 9.11, so that a primary packing leak can be detected.

9.6 Manufacturing in China

Two pump-manufacturing companies were visited in China in 2000 to evaluate their product line and production facilities. One of the companies built centrifugal pumps and dabbled in positive displacement pumps for limited markets, and the other company was seriously pursuing the manufacturer of products, which included positive displacement pumps with pressures up to 7,500 psi and 200 hp. Both companies built manufacturing plants just outside of crowded cities. These plants

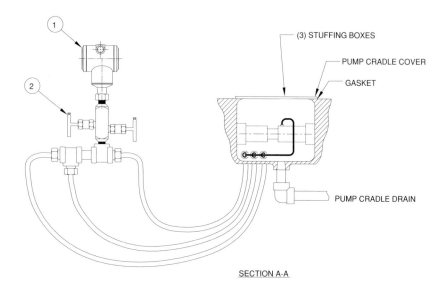

Figure 9.10. Pollution prevention system using *Smart Hart*

seemed like facilities that could be found in the United States back in the 1950s—the facilities emitted pollution and had somewhat primitive working conditions. The machine tools were old, and the production methods were not the most modern.

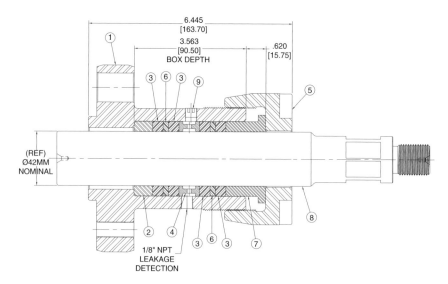

Figure 9.11. Pollution prevention using double packing for triplex pump

By 2004, the new plant facilities that were started in 2000 had been outgrown, and the companies were building more modern shops to add more capacity. The product quality had improved, and the size range of positive displacement pumps had increased to include 600-hp pumps. To improve the pumps' casting quality and prevent causing local pollution, a remotely located foundry was selected to produce one company's power frame castings. The remote foundry, which was formally owned by the Chinese government, became privately owned, and it grew in size and production capacity. Even though the foundry was located in a remote area of China, the company was concerned about pollution and was well underway on the construction of a new plant farther out in the country.

The conditions in these pump factories in 2004 were reminiscent of those in the United States in the 1960s and 1970s, when companies started considering the effects of pollution on the environment and when production facilities began to use more sophisticated machine tools and manufacturing methods. With the rate of improvement, the factories and foundry visited in China will soon resemble U.S. facilities after they were modernized.

9.7 Safety Guidelines

The WaterJet Technology Association (WJTA) has improved safety worldwide in the water-jetting industry. Some of its earliest work focused on outlining safe work practices that described the basic practices for using high-pressure water jets when cleaning or cutting materials. These recommended practices apply to the operation of all types of high-pressure water jets that are normally used in construction, maintenance, repair, cleaning, cutting, and demolition work. The practices describe types of water-jetting equipment and include recommendations about how to run the equipment and how to train equipment operators. In addition to the WJTA's guidelines, present or proposed industry standards exist. These include those of the U.S. Occupational Safety and Health Administration (OSHA), the American Society of Testing Materials (ASTM), and the American National Standards Institute (ANSI). Equipment manufacturers' recommendations are also included in standards when appropriate.

9.7.1 Correct Operation

During any water-jet activity, the correct operation and application of the water-jet equipment are the operator's responsibilities; the operator should be familiar with the identification of high-pressure pumps, high-pressure fittings, hoses, guns, and accessories. The modification of water-jet equipment or accessories is not recommended without the prior written approval of the manufacturer of the equipment. Serious harm or injury may result from the misuse of water-jet equipment or the use of improper fittings, hoses, or attachments. The WJTA does not guarantee that the practices described and the recommendations contained in its pamphlet of recommended practices will prevent harm or injury.

9.7.2   Correct Response

A person injured by being hit with a water jet will not necessarily see the full extent of the injury, particularly the internal damage and depth of penetration. Even though the surface wound may be small and may not even bleed, it is quite possible that large quantities of water may have entered the skin, flesh, and internal organs through a very small puncture. The spread of microorganisms through a wound of this type is a very real concern; the injury should, therefore, be carefully monitored for several days. In the event of an injury, immediately take the injured person to a hospital and inform the doctor of the cause of the injury. To ensure that the doctor knows and understands the cause of the injury, all water-jet operators should carry a waterproof card that is easily accessible and that outlines the possible nature of the injury. The wording on the card might be as follows:

> This person has been involved with high-pressure water jets at pressures up to 55,000 psi, (375 MPa, 3750 bar), with a velocity of up to 2000 mph (3,300 kph). Please take this into account when making your diagnosis. Unusual infections have been reported with microaerophilic organisms that tolerate low temperatures. These organisms may be gram-negative pathogens, such as those found in sewage. Bacterial swabs and blood cultures may therefore be helpful.

9.7.3   Personal Protection

All personnel should follow the OSHA regulations for wearing personal protective equipment, including head protection, eye protection, hearing protection, body protection, hand protection, foot protection, and respirator protection. Only personnel who have undergone a proper training program and who have demonstrated the knowledge and skill as well as gained the experience to perform all likely assigned tasks shall operate high-pressure water-jet equipment.

For operating a handheld gun, the back thrust from a jet can be calculated from the following equation:

$$Back\ thrust\ (pounds) = .05266 \times Q\sqrt{\Delta P}.$$

In this equation, Q is the flow rate in U.S. gallons per minute, and P is the jet pressure measured in pounds per square inch. For example, an operator working with a jet producing 10,000 psi at 10 gpm will experience a force equal to 52 lb, calculated from $0.052 \times \sqrt{10,000}$ psi $\times$ 10 gpm. It is not recommended that any one person be required to withstand a back thrust of more than one third of his or her body weight for an extended period of time. For this latter example, this means that the operator should weigh at least 156 lb to operate the nozzle.

9.8   Glossary of Water-Jetting Terms

The definitions of and guidelines for operating components of water-jet systems have been outlined because similar equipment is used for water-jet operations in many countries. When possible, it is considered important that the same words be used for similar parts of a system. For this reason, many of the following terms have been based on original definitions from the WJTA and AHPWC organizations.

9.8.1   Abrasives

Abrasives are solid particles, either soluble or insoluble in carrier fluid, that are introduced into a water jet before it hits the target surface. Such particles are often used to increase the effectiveness of pure water jets for some applications. Abrasives can be used to prepare a surface for painting as well as for cutting materials.

9.8.1.1   Abrasive Feed System

An *abrasive feed system* includes a storage vessel or hopper for the abrasive, a hose or tube to carry the abrasive to the point where it is inserted into the water jet, and a device for inserting the abrasive into the water-jet stream.

9.8.1.2 Abrasive Jet

A water jet that requires solid particles to be introduced into the jet stream before the jet hits the target is called an *abrasive jet*. There are three types of abrasives that can be used and they differ according to how they are introduced; these are the entrained abrasive, external abrasive, and slurry abrasive.

9.8.1.3 Entrained Abrasive

An *entrained abrasive* refers to particles that are added to the jet stream after the jet has accelerated through an orifice but before the resulting stream has been reshaped through a collimating nozzle.

9.8.1.4 External Abrasive

For an *external abrasive*, the particles are added to the jet steam after it has left the final orifice.

9.8.1.5 Slurry Abrasive

A *slurry abrasive* refers to particles that are added to the water before it is accelerated through an orifice.

9.8.2 Automatic Pressure-Relief Devices

A high-pressure water-jet system provides a way of automatically limiting the system pressure through the use of *automatic pressure-relief devices*. Several types of devices exist:

- Automatic pressure-regulating valves
- Bursting and rupture disks, when set in a proper holder
- Bypass valves
- Pressure-relief valves

An automatic pressure-relief device should be mounted close to the discharge outlet of the pressurizing pump because the pressure at this point is the highest in the system. This location will also allow a more immediate reduction in pressure, without retaining higher pressures in downstream components of the system.

9.8.2.1 Automatic Pressure-Regulating Valve

A valve used to automatically control the working pressure in the high-pressure water-jet system by controlling the bypassing water flow is termed an *automatic pressure-regulating valve*. When the pressure in the system exceeds a set level, the valve will partially open. As the valve continues to open, more water is bypassed and less flows to the nozzle. The water passing through the valve can be directed either back to the pump's supply reservoir pump or to another disposal. An automatic pressure-regulating valve may be used to control the system's operating pressure, and, if so, the valve should be checked to ensure that it is set at the correct value before it is used in each water-jet operation. When there is no demand for high-pressure water, this valve may be used to ensure that the system pressure is brought down to a low level; hence, the valve is sometimes referred to as an *unloading valve*.

9.8.2.2 Bursting or Rupture Disk

A *bursting* or *rupture disk* is normally a metal disk held in a specially designed holder that will fail when the pressure applied to it exceeds a set level. Disks can be made of different materials and are of different sizes. A disk of the proper size should be used for a given operating pressure. The holder should be designed and located so that any water passing through it is not directed at an operator or other component of the high-pressure water-jet system.

9.8.2.3 Bypass Valve

A *bypass valve* can be adjusted by the operator, either manually or automatically, to control the flow and the pressure of the jet stream issued from the nozzle.

9.8.2.4 Pressure-Relief Valve

A *pressure-relief valve* is normally held in the closed position by a mechanical device, such as a spring. It is designed to open when the pressure in the system exceeds a set value.

9.8.3 Burst Pressure

The *burst pressure* is the internal pressure of a high-pressure water-jet system component that will cause that component to fail. It is important to note that high-pressure equipment undergoes cyclic loading because of the reciprocating movement of plungers or pistons. This loading will fatigue the parts of the system so that, with time, the strength of the components will decline.

9.8.4 Catcher

When a plain or abrasive laden water jet is used in a cutting operation, a device called a *catcher* can be placed on the opposite side of the workpiece to catch the spent jet, abrasive, and particles of the material. This catcher is fitted with a waste tube that carries this spent material away from the area.

9.8.5 Changeover Valve

A valve that the operator can adjust to send the water from the pressurizing pump to either one or several pieces of water-jet equipment supplied by the pump is termed a *changeover valve*. This valve can be operated either manually or by a secondary power circuit attached to the high-pressure water-jet system.

9.8.6 Collimating Nozzle

A *collimating nozzle* is a secondary nozzle used below the mixing chamber to refocus the stream of high-pressure water and abrasive in conventionally mixed abrasive water-jet systems.

9.8.7 Dry Shutoff Control Valve

A valve that is normally manually controlled by the lance or nozzle operator to start and stop water flow to the nozzle is called a *dry shutoff control valve*. Although closing this valve stops the water flow to the nozzle, it keeps the pressure level in the supply line at the system's working pressure. When this valve is used, the system should also be fitted with an automatic pressure-regulating valve to ensure that the system's working pressure is not exceeded. When this valve is used, the operator should release the pressure in the water supply lines after the pump has been shut down. This act will ensure that the system is not left under pressure. The valve can also be operated by a secondary power circuit attached to the high-pressure water-jet system.

9.8.8　Double-Trigger Gun

A *double-trigger gun* requires that two triggers be activated by the operator, one with each hand, to generate a high-pressure water jet.

9.8.9　Dump Systems

A *dump system* should be equipped with a device that will shut down the pump, cause it to idle at low revolutions per minute, make it bypass the flow, or cause it to reduce the discharge pressure to a low level. Only the nozzle operator should manually control the dump system. The dump system actuator device should be shielded to prevent accidental operation. This device should be controlled by the operator's hand or foot and should dump the high-pressure water stream if the operator releases it. If the water was dumped but was not immediately released to the open air and instead passed into a dump line, the line must be secured so that it does not whip when active.

9.8.9.1　Dump Control Valve

The operator of the lance or jetting equipment normally manually controls the *dump control valve*. The operator normally closes this valve so that the water can be sent to the nozzle. When the valve is released, it will automatically stop water flow to the lance and/or nozzle assembly because it opens a much larger flow passage through which the water is diverted at low pressure. For this process to be effective, both the passage through the valve and the diameter of the relief line should be large enough so that no significant resistance to the water flow develops, even at maximum pump output. A valve size should be selected that will not cause generation of significant back pressure at the maximum possible pumping rate of the pump. This valve can also be used with an electrical or pilot pressure system that includes additional circuits that must be engaged for the valve to actuate. These systems should be designed such that if the valve fails, it opens.

9.8.9.2　Solenoid and Electrically Operated Control Dump Systems

All *solenoid* or *electrically controlled dump systems* should be of fail-safe design. Voltage of an alternating current (AC) or direct current (DC) dump system, handled by personnel, should not exceed 24 V and should be fuse protected.

9.8.10   End Fittings and Couplings

High-pressure hose-*end fittings* and *couplings* should be manufactured to be compatible with the hose and should be tested as a unit.

9.8.11   Filter or Strainer

The water system should be equipped with a *filter* or *strainer* to prevent particles from restricting the flow through orifices in the nozzle or damaging seals of the pump. The strainer or filter should be capable of removing particles smaller in size than half the diameter of the smallest opening. Smaller filter sizes are strongly recommended because the pump and other system components will last much longer.

9.8.12   Flexible Lance

A flexible tube or hose section carrying water to the nozzle is called a *flexible lance*; this lance is normally located between the trigger or control valve and the nozzle.

9.8.13   Foot-Controlled Valve

A *foot-controlled valve* is designed so that the operator can activate the valve by using his or her foot. These types of valves allow an operator to use both hands to hold and move the lance and/or nozzle assembly. When a foot valve is used, it must be placed within a frame that will guard the valve from being accidentally operated; water connections must also be sturdy enough so that the valve will not accidentally be moved or knocked over when it is used.

9.8.14   High-Pressure Hose

A flexible hose that can be used to carry water and/or other fluids from one part of the high-pressure water-jet system to another is termed a *high-pressure hose*. The hose should have a burst rating of a minimum of 2.5 times the intended working pressure certified by the manufacturer. The high-pressure hose should be tested at 1.5 times the working pressure. It is important to note that the hose should not be used at a pressure above the manufacturer's recommended working pressure.

9.8.15 Fiber-Reinforced Hoses

The failure of a *fiber-reinforced hose* may cause a "pinprick" hole and the possible formation of a hazardous jet. Adequate safety precautions must be taken.

9.8.16 High-Pressure Water-Jet Systems

High-pressure water-jetting systems are water delivery systems with nozzles that increase the speed of liquids. Solid particles or additional chemicals may also be introduced, but the exit in all cases will be a free stream. In terms of the field's recommended practices, the system should include the pumps (pressure-producing devices), hoses, lances, nozzles, valves, safety devices, and attached heating elements or injection systems. High-pressure water jets are used at several ranges of pressure. The following divisions are made to clarify these ranges.

9.8.16.1 High-Pressure Water Cleaning

High-pressure water cleaning refers to the use of high-pressure water, with or without the addition of other liquids or solid particles, to remove unwanted matter from various surfaces; the pump pressure is usually between 5,000 psi (340 bar) and 30,000 psi (2041 bar). When the term *high pressure* is used without further qualification, it refers to jets being used at pressures below 30,000 psi (2041 bar).

9.8.16.2 High-Pressure Water Cutting

High-pressure water cutting refers to the use of high-pressure water, with or without the addition of other liquids or solid particles, to penetrate the surface of a material for the purpose of cutting that material; the pump pressure is usually between 5,000 psi (340 bar) and 30,000 psi (2041 bar). When the term *high pressure* is used without further qualification, it refers to jets being used at pressures below 30,000 psi (2041 bar).

9.8.16.3 Pressure Cleaning

The use of pressurized water, with or without the addition of other liquids or solid particles, to remove unwanted matter from various surfaces is called *pressure cleaning*; the pump pressure is below 5,000 psi (340 bar).

9.8.16.4 Pressure Cutting

The use of pressurized water, with or without the addition of other liquids or solid particles, to penetrate the surface of a material for the purpose of cutting that material is termed *pressure cutting*; the pump pressure is below 5,000 psi (340 bar).

9.8.16.5 Ultra-High-Pressure Water Cleaning

Ultra-high-pressure (UHP) *water cleaning* refers to the use of high-pressure water, with or without the addition of other liquids or solid particles, to remove unwanted matter from various surfaces; the pump pressure exceeds 30,000 psi (2041 bar).

9.8.16.6 UHP Water Cutting

UHP water cutting refers to the use of high-pressure water, with or without the addition of other liquids or solid particles, to penetrate the surface of a material for the purpose of cutting that material; the pump pressure exceeds 30,000 psi (2041 bar).

9.8.17 Hose Assembly

A *hose assembly* refers to a hose with a suitable end coupling attached, at each end of the hose, in accordance with the manufacturer's specifications.

9.8.18 Hose Shroud

A length of flexible material usually formed into a tube around a hose end coupling or across the connection to the jetting gun is termed a *hose shroud*. The shroud provides some instantaneous protection should a hose burst. It will not form a permanent barrier to the flow of water from a damaged hose.

9.8.19   Jetting Gun

The hand-operated device that is often used in manual water jet is called a *jetting gun*. This gun is normally connected to the high-pressure system by a high-pressure hose assembly. The gun is made up of a control valve (mounted within a guard), a lance section, and a nozzle assembly, which may include one or more nozzles. The gun may also include a support bracket and shoulder pad and/or one or more support handles. The gun can be further defined by the type of control valve that is used to release the pressure. If the pressure is dumped to the atmosphere when the valve is released, then the gun is a dump gun; if the pressure is retained in the system by using a dry shutoff control valve, then the gun is a dry shutoff gun.

9.8.19.1   Jetting Gun Extension

The *jetting gun extension* is a length of tubing, either made from piping or lances, that is used to extend the reach of the gun. Extension pieces should be manufactured from suitable material and with proper end connections for the application. Each extension should have a minimum burst strength of at least 2.5 times the highest actual working pressure used.

9.8.19.2   Jetting Gun Trigger

The control valve has a *jetting gun trigger* that makes it easier for the operator to control the device. This lever, or trigger, should be designed for easy operation by an operator wearing gloves. The trigger should include a catch or other method of lockout so that it cannot be operated until this catch is released.

9.8.19.3   Jetting Manifold

The *jetting manifold* provides an attachment at the end of the lance into which individual nozzles or nozzle holders may be threaded to distribute the water jets over a given pattern. Alternating nozzles may be directed forward and backward from the manifold to reduce or even balance the thrust exerted by the jets on the manifold. The lance and operator should be so arranged to shut the system down without damage if a nozzle is plugged, which would cause unbalanced thrust.

9.8.20 Lancing

Lancing is an application whereby a rigid or flexible lance and nozzle combination is inserted into, and retracted from, the interior of a pipe or tubular product.

9.8.20.1 Flexible Lance

A flexible tube or hose section carrying water to the nozzle or nozzle manifold is called a *flexible lance.*

9.8.20.2 Rigid Lance

A *rigid lance* is a tube carrying water to the nozzle or nozzle manifold.

9.8.21 Moleing

Moleing is an application whereby a hose fitted with a nozzle is inserted into, and retracted from, the interior of a tube. It is a system commonly used with a self-propelling nozzle for cleaning the internal surfaces of pipes or drains. Moles can be self-propelled by their backward-directed jets or can be manufactured to be fitted with various shapes, sizes, and combinations of forward-directed and backward-directed jets. A mole should include a section of rigid pipe or tubing (located directly behind the nozzle assembly) that is long enough to prevent the mole from turning around in the pipe.

9.8.22 Nozzle

A device with one or more orifices through which the water discharges from the system is called a *nozzle*. The nozzle restricts the area of fluid flow, accelerating the water to the required velocity and shaping it to the required flow pattern. Nozzles are also commonly referred to as *bits*, *tips*, or *orifices*. A nozzle may be further defined by the type of jet that it produces.

9.8.22.1 Nozzle Holder

The threaded fitting that holds a nozzle insert and attaches it to the jetting manifold or shotgun lance extension is called a *nozzle holder.*

9.8.22.2 Nozzle Insert

A replaceable nozzle usually fitted with one orifice and designed to fit into a nozzle holder is a *nozzle insert.*

9.8.23 Operator

An *operator* is person who has been trained and has demonstrated the knowledge, skill, and experience to assemble, operate, and maintain a water-jet system.

9.8.24 Operator Trainee

An *operator trainee* is a person who is not qualified, because of lack of knowledge, skill, and/or experience, to perform as an operator without supervision.

9.8.25 Orifice

An *orifice* is the opening at the end of a nozzle through which the water or fluid jet exits from the system.

9.8.26 Pressure

9.8.26.1 Pressure Gauge

A high-pressure water-jet system should be equipped with a *pressure gauge* that indicates the pressure of the system when it is performing a job. Gauges shall have a scale range of at least 50% above the maximum working pressure of the system and should be fitted with a pressure snubber for more accurate pressure reading.

9.8.26.2 Pressure Intensifier

A *pressure intensifier* is a pump that increases the pressure of water supplied to it by using the reduction in area of a common piston and multiplying the pressure from the driving fluid, which is usually oil.

9.8.26.3   Pressure Pump

A *pressure pump* will increase the pressure of water delivered to it and deliver the water into a common manifold to which either flexible hoses or rigid tubing (connecting to lances and nozzles) is attached. These pumps can be either mobile or permanently mounted and are most often of a positive displacement plunger-style that will provide a constant flow of water at a given speed of rotation. The pump should have a permanently mounted tag designed to provide the following information:

- Name of manufacturer
- Model, serial number, and year of manufacture
- Maximum performance in terms of gallons or liters per minute and pressure in pounds per square inch (or bar)
- An outline of recommended safety procedures

9.8.26.4   Pressure Relief

The high-pressure water-jet system should be equipped with an automatic relief device, called a *pressure relief*, on the discharge side of the pump.

9.8.27   Pulsating Water Jet

A jet that consists of individual slugs of water or liquid is called a *pulsating water jet*.

9.8.28   Rigid Lance

A rigid metal tube used to extend the nozzle from the end of the hose or jetting gun is called a *rigid lance*, as mentioned in section 9.8.20.2.

9.8.28.1   Rigid Lancing

Rigid lancing is an application whereby a lance or jetting gun extension is fitted with a nozzle, nozzle assembly, or nozzle manifold. This lance is inserted into, and retracted from, the interior of a tube, tank, or vessel.

9.8.29 Self-Rotating Nozzle Assembly

A *self-rotating nozzle assembly* includes a bearing or swivel assembly that fits onto a hose or lance section. The device contains at least two jets that are offset so that the reaction force from the jets causes the nozzle assembly to rotate without any additional external force being applied.

9.8.30 Shotgunning

Shotgunning is a handheld operation in which an assembly of a lance and a nozzle can be manually manipulated in virtually all planes of operation.

9.8.31 Spray Bar

A *spray bar* is a special manifold designed to distribute nozzles along a linear tube or pipe. This bar is most often used to provide an array of fan jets that overlap, and it is frequently used to clean large areas.

References

Association of High-Pressure Water-Jetting Contractors in the United Kingdom. *Code of practice for the use of high-pressure water-jetting equipment*. Association of High-Pressure Water-Jetting Contractors in the United Kingdom.

Easterbrook, G. (2005). The good earth. *Reader's Digest*. Pleasantville, NY: .

Gracey, M. T. (1992). Ecological solutions used for high-pressure water-blasting in the petrochemical industry. *The Mexico International Oil, Gas, & Petro Chemical Equipment Technical Conference*, Mexico City Mexico.

Gracey, M. T. (2001). Hydrobalanced packing system for high-pressure pumps. *American Water Jet Conference*, Minneapolis, Minnesota.

Gracey, M .T., and Palmour, H. H. (2000). Artificial lift hydraulic improvement. *Southwestern Petroleum Short Course*, Lubbock, Texas.

WaterJet Technology Association (WJTA). *Recommended practices for the use of manually operated high-pressure water-jetting equipment*. WJTA.

Chapter 10

Hot Water Washdown Unit

10.0 Pumps Handling Hot Water

Hot-water pressure washers are a common part of the small positive displacement pump market. Pressures usually run in the 1,000–3,000 psi and may incorporate an onboard heater fired by fuel oil. When it comes to offshore hot-water pressure washers, electric heaters are often used to elevate the temperature of the water supply to the positive displacement pump. These electric, hot-water pressure washers are often rental units that are compact and portable, but the new offshore complexes are being designed with cleaning systems that are permanently installed within the washers.

In October 2001, ABB Lummus received an award to supply a high-pressure, hot-water washdown unit to Exxon-Mobil for its offshore platform, named Kizomba A (built for Angola). The offshore platform was manufactured according to the exact specifications of Brown &

Root, a well-known engineering group. The unit was to provide 1,800 psi at 40 gpm and 150° F water to four work stations on Kizomba A.

In this chapter, the changing rules of agreements between pump suppliers and oil companies will be discussed as they apply to this water-jetting package originally built in Houston, TX. The documents, drawings, electronic submittals, approval cycle, inspection, and final acceptance testing required for the project proved to double the scope of work for the equipment. The project reveals a new way of doing business with major oil companies, including an increase in the scope of engineering and a large amount of documentation to complete each job. This chapter explains how 95 submittals were required and how the number of submittals affected the cost and schedule for building a pump system.

The increased amount of work also extended to subsequent phases of the project. In addition, unusual aspects, such as computer-controlled jetting pressure and water heating, complicated the project. With other oil companies demanding similar specifications for their equipment, it is important for a pump supplier to acknowledge this trend. Suppliers have to understand the steps involved if they are interested in working with major oil companies.

10.1 Pump Unit Engineering

For pump suppliers, there seems to be a new way of doing business with major oil companies. The changes may have developed because of an increase in the number of regulatory bodies and the high cost of insurance, or it may have come from financial or safety concerns. Whatever the reason, engineering to be done by the vendor and the increased documentation requirements must be considered during the bidding process.

When Weatherford received a request for a bid in July 2001, the company believed the request was straightforward when compared to past and present job specifications even though the list of technical information and submittals was a little long. In October of that same year, the agreement was reached with ABB Lummus to provide the hot-water, high-pressure washdown unit to Exxon-Mobil. Figure 10.1 shows the completed pump unit, and Figure 10.2 provides a closer view of the high-pressure, positive displacement pump selected to meet the specifications.

Over time, Weatherford realized that the project was anything but straightforward:

Figure 10.1. Hot water washdown unit

- The purchase order was let before the work was fully defined.
- A large number of submittals were due in 2–4 weeks after the order date.
- Questions could not be answered about system details.
- The kick-off meeting took 2 months to schedule.
- Changes were requested that took additional time to submit.
- The prime contractor, main subcontractor, and component vendors were overwhelmed by the demands of the job documentation.

During the quotation stage for this high-pressure pump unit, the experienced personnel who were involved made suggestions to the engineering group/oil company customer, but few of them were accepted. It was suggested that instead of supplying hot water to the pump suction, the water could be heated after the pump to improve maintenance cost. This change would have also prevented hot water from being dumped by the temperature relief valve, as shown in Fig. 10.3, so the heater power consumption would decrease dramatically. Another suggestion was to place the heater control panel in the

Figure 10.2. Triplex pump used in the hot water washdown unit

customer's control room so that the large explosion–proof cabinet, as shown in Fig. 10.4, could be eliminated.

The detergent injection system that was specified could supply soap to the four-gun operation; however, when one gun was to be operated, the eductor, as shown in Fig. 10.5, would not have enough of a pressure drop across it to draw the detergent. Other methods of detergent injection were proposed, but these ideas were also rejected. The suggestions to improve the design and operation were dismissed or ignored.

10.2 High-Pressure Pump Experience

Hot water is being used in areas such as oil and rubber removal, including as in the removal of rubber from airport runways. Hot water has also long been used in the pressure washer industry in combination with steam and detergents.

Before accepting this project, both the prime contractor and the subcontractor for the ABB Lummus project had experience in

Figure 10.3. Air-operated high temperature relief valve

manufacturing pump units, hot-water wash units, and high-pressure water-jetting equipment. The sub contractor, ACME Cleaning has been building hot-water washdown units for another oil company for the last several years and has delivered a dozen electric-heated, motor-driven, and skid-mounted units. The prime contractor, Weatherford, is known for building high-pressure pumps and packages for a variety of pumping requirements, including water jetting. Additional experience has come from using ultra-high-pressure (UHP) water jetting has been used to remove existing coatings from offshore rigs, and low-pressure water jets for multiple purposes. Oil companies have many uses for the pumps they purchase; past projects have involved methanol injection, restart pumping, petrochemical product applications, and special chemical handling requirements. For most contracts and projects, the specifications for the equipment are followed, and documentation is provided at the end of the job as required by the purchase order. Completion of the job involves approval drawings, material certifications, and operational instruction manuals. Welding procedures, weldor qualification, and inspections are a normal part of the process also.

Figure 10.4. Explosion-proof 450 KW heater control panel

10.3 Project Delays

Normal reasons exist for the delay of progress on a job. These include taking time for vacations, holidays, and other projects; a lack of personnel to do the job; and the time needed for personnel to become familiar with the requirements of the job. Time is required for engineering, system drawings, and customer approvals; however, this part of the process is over after a few weeks.

In the case of the ABB Lummus project for Exxon-Mobil, delays were due to some unusual aspects. Each submittal had to be sent to ABB Lummus electronically and accompanied by certain forms. It took about 4 weeks to get a response; when the document was finally returned, a "status" was assigned to the document (A1, A2, N1, N2, etc.). Each of the codes had an associated meaning; for example, N2 meant "Comments as noted. Proceed with fabrication in accordance with comments, and resubmit corrected drawings as 'final certified' within 2 weeks or sooner."

Figure 10.5. Detergent injection system

The obstacle in this approval method is that subsequent submittals of each document were returned with additional comments, so the submittals were never really approved. When the components were ordered or the item built, it was still subject to change. For example, when the project finally started, the customer's first expeditor called to say that a list of required documents needed to be immediately submitted. These documents and drawings were submitted for approval. Weeks later, when one of these documents was returned with an N2 status, for example, the second expeditor insisted that the item be ordered or built because of the N2 status. The next time through, this same document could have come back with more notes or it may have been rejected, so the wrong thing was ordered or fabricated.

When an inefficient process such as the one just described is repeated 95 times, serious problems arise. Extra costs are incurred, and delays are never-ending. The expeditors were calling the subcontractors and vendors to push for delivery, and, meanwhile, answers could not be obtained about the specification requirements. Meetings were scheduled that seemed to raise more questions and bring more personnel into the loop. Inspectors would give directions to get the job moving, but they were later countermanded by the next set of inspectors. If 95 documents could be submitted once or maybe twice, the paperwork process would have been more reasonable, but one document (e.g., set of drawings, engineering calculations, or forms) would go through the system multiple times (up to five to six times) even after the equipment had been shipped.

10.4   Overcoming Obstacles

In more than one meeting, ABB Lummus offered to end work on the project so that the customer could find another contractor—the obstacles appeared insurmountable. Each time this happened, however, Exxon-Mobil personnel promised that the company would help ABB Lummus surmount the hurdles. There was even an attempt on the engineering company's part to help define the requirements, but this company was in the same boat and had problems of its own.

To provide a solution for the impact of the large amount of documentation on the schedule, a person was hired for document control. This project manager took over communication with the expeditors and inspectors. The project engineer and the main subcontractor negotiated the first change order and worked on the subsequent change orders, while defining the customer's requirements until the project manager could take over the financial negotiations. Vendors were contacted on a regular basis for updates on documents and drawings until the information was received for submittal or resubmitted.

The fact that there was no penalty for late delivery of a purchase order was one of the only redeeming factors about the job. It was also made clear in the quotation and in the acceptance letter that items such as positive material identification (PMI) and full-system tests were not included, and a statement was even received that the full test was not required. These were unknown cost and would have to be quoted, if needed. The obstacles were gradually overcome to complete the system. Figure 10.6 shows the piping and instrumentation diagram (P&ID) of the skid-mounted offshore package with heater, tanks, and stainless steel piping in the final stage of completion.

10.5   Final Acceptance Test

Quality control plans, test procedures, and numerous inspections were part of the project. The intent was to run the system cold and

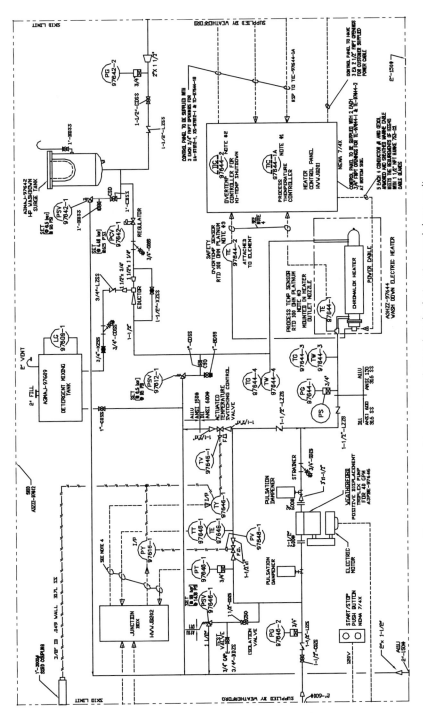

Figure 10.6. Portable generator for testing the hot water washdown unit

Figure 10.7. Photo of generator used for testing

show flow and pressure because the heater required 450 kW, and it was not available in the facilities. Later in the project, it was decided that a full functional test was required after all, and there was more negotiation about the manpower and electrical needs. A large generator was rented, and the necessary manpower was provided to run the pump, heater, controls, and four pressure-washer guns. This process included the use of a water tank, supply hoses, and a charge pump, but the most difficult part of the process was to provide a simulation of the customer's control of the discharge pressure. Figure 10.7 shows the 450-kW generator required to power the water heater and run the system. The numerous people involved in the final acceptance of the equipment included inspectors, customer representatives, and engineers.

10.6 Conclusions About the Hot Water Unit

The new way of doing business for some major oil companies increases the scope of engineering and documentation. A product such as a high-pressure, hot-water washdown unit may be fairly simple to engineer, but the purchase order may include specifications that the contractor should fully understand before accepting the project. Some engineering companies and their customers, the major oil companies, are now enforcing what was once "boiler plate" for reasons of liability or safety. The lessons learned from the ABB Lummus and Exxon-Mobile project summarized in this chapter include (a continuation of the summary on the project can be found in Dai and Gracey [2005]):

1. The primary cause of delivery delays stemmed from the drawing and document approval process. Delivery schedules should start from the date that all documents have been fully approved.
2. One of the greatest delays in the process was caused by the time it took to submit, reply to, and resubmit documents. All notes and changes should be finalized on the first submittal of a document.
3. Extra time should be allowed for vendors to furnish documentation that is part of the submittal process. The vendor may have to develop the information, and it may have to be reworked before approval.
4. Progress payments should be required especially if there are delays in the project completion.
5. All commercial issues should be finalized before the equipment is released for shipment.
6. Some requirements cannot be met without the control equipment that is to be furnished by others. The final acceptance test (FAT) should be studied carefully so that exceptions can be noted.

References

Dai, W., and Gracey, M. T. (2005). Offshore washdown pump unit, a manufacturing case study. *WJTA American Water Jet Conference, Houston, Texas.*

Gracey, M. (2001). Where the rubber meets the road. *Cleaner Times.* Little Rock, AR.

Gracey, M. (1997). Unusual uses for high-pressure jetting. *Cleaner Times.* Little Rock, AR.

Gracey, M. T., and Berry, R. O., Jr. (2003). Manufacturing case study involving major oil company. *WJTA American Water Jet Conference, Houston, Texas.*

Chapter 11

Troubleshooting High-Pressure Pumps

11.0   Introduction

High-pressure pumps and systems provide a variety of opportunities to use engineering skills and past experience to solve the problems that arise during start-up, testing, and daily operation. Proper installation is of prime importance because the fluid-lifting capabilities of a reciprocating pump are limited. As minimum requirements, the suction should be flooded, and a positive suction head for satisfactory operation should be installed. Improper installation is often the source of the initial problems encountered or the source of persistent operational complaints. These types of requirements and scenarios should be

167

considered when installing or troubleshooting a high-pressure pump or system.

11.1   Suction Lines

The suction line should be at least as large as the pump suction—one size larger is preferable. It is advisable to use a flexible hose rather than hard piping when possible. The number of ells, turns, and restrictions should be kept to a minimum, and an ideal installation would not have any of these possible sources of complication. If ells are necessary, they should not be closer than 5 ft from the pump inlet, and no two ells should be closer together than 3 ft. Branch laterals and 45-degree ells are preferred in a suction line rather than 90-degree ells. All valves in the suction line should be a full, open type.

11.2   Discharge Piping

Discharge piping should extend at least 5 ft from the pump before any turns or ells are installed, and flexible hose is recommended to eliminate line stresses and to dampen vibration. Any ell should be a 45-degree type or lateral, and two 45-degree ells are preferable to a single 90-degree ell.

11.3   Relief Valves

Relief valves or rupture disks help protect pumps from failure due to excessive pressure and must be installed in the discharge piping before any control valves are installed. Operation without a properly set and installed relief valve is dangerous and can cause loss of life and damage to the high-pressure system.

11.4   Start-Up

To start the high-pressure pump, open the suction and discharge line valves, and roll the pump over by hand to permit fluid to fill the fluid cylinder. If rolling the pump by hand is not possible, bump the motor or engine clutch to make the pump turn momentarily. If possible, bleed all the air from the system by removing the discharge hose or disconnecting the discharge line. Trapped air can cause the pump to run in a rough manner; therefore, before running the pump continuously, eliminate trapped air and ensure that the high-pressure pump is getting sufficient fluid.

11.5 Pulse Dampeners

A pulsation dampener is a worthwhile investment when it is properly selected, correctly precharged, and located near the pump. It can provide longer valve and spring life while reducing the surges. A dampener can promote smoother operation of the pump when used in the suction line and can reduce pulsation when used in the discharge line.

A simple setup for a water-blasting pump is shown in Fig. 11.1. A pressurized supply line pushes the water through the filter and into a vertical tank. Air is pushed to the top of the vertical tank to help degassify the water before it is supplied to the high-pressure pump. The discharge from the pump is usually through high-pressure hose, which tends to absorb the pulsation for pumps in the 10,000-psi range. For newer ultra-high-pressure (UHP) pumps, accumulators are often used to smooth the flow.

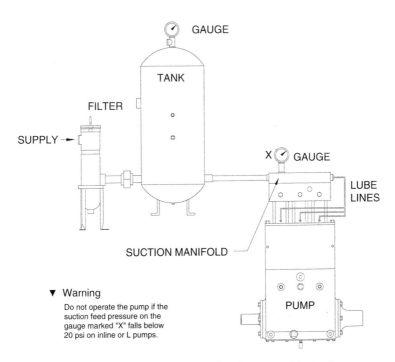

Figure 11.1. Suggested inlet piping for a water blast unit

11.6   Final Acceptance Test

A final acceptance test (FAT) allows users to discover problems with a high-pressure pump system. For example, the results of the FAT for the system described in Chapter 10 showed that the first unit was performing as expected. This particular pump was assembled in Texas and delivered the 40 gpm at 1,800 psi, as required, and delivered hot water to four shut-off style guns during the FAT.

The second unit in the series was assembled in Louisiana, but when it came time for the FAT, the pump did not perform well—the testers suspected that there was not enough water reaching the suction side of the pump. The project manager, the customer's inspector, and the shop people were waiting and prepared to address any problems that arose during the FAT. A simple and fast method to analyze the pressure drop from the 70-psi inlet pressure through the piping, valves, and tanks had to be used. Analyzing the resistance of valves and fittings to flow of fluids seemed to be the quickest way to understand the source of the problem. From the P&ID of the system shown in Fig. 10.6, a list was made of the elements in the plumbing leading to the positive displacement triplex pump. This list is shown in Table I.

As determined by the charts in Appendix D, the pressure drop for 1–½-in pipe is around 4.78 psi per 100 ft. Multiplying 4.97 × 4.78 = 23.76-psi pressure drop in the water inlet side of the plumbing system. If the incoming water is adjusted to 70 psi at 40 gpm, then 70 psi – 23.76 psi pressure drop = 46.24 psi, which should be measurable at the pump inlet. To solve the problem in the second unit for the system described in Chapter 10, gauges were placed at the pump and various places in the plumbing system until the item, which caused the most pressure drop, was located. The check valves were specified as *API #12 Ball Lift Type*, but plug-type, standard-port check valves were purchased and installed in the system. When the check valves were replaced, the pressure drop problem was eliminated.

11.7   Start-Up Problems

To explain start-up problems with a high-pressure pump unit, this section provides an example. Two offshore wastewater pump units were built to provide a flow of 8 gpm at 6,000 psi; during start-up before the final tests, the belts squealed when the first pump was loaded.

Table I
Resistance to Flow in Terms of Equivalent Pipe Length

Element in the System	Quantity	Equivalent Pipe Length
Pipe, schedule 80, 1–½ in	45.5 ft	45.5
Elbow, 1–½ in	17 × 4 in of pipe	68
Valve, check 1–½ in	2 × 40 in of pipe	80
Valve, globe 1–½ in	1 × 40 in of pipe	40
Thermoweld probe	2 × 40 in of pipe	80
Eductor, 1–½ in	1 when in bypass	0
Y-strainer, 1–½ in	2 when clean	80
Union, 1–½ in	7 full bore	0
Flange, 1–½ in	6 full bore	0
Swage, 2 × 1–½ in	1	2.5
Tee, 1–½ in	9	81
Valve, 1–½-in ball	1	1
Swage, 1–½ × 1–¼ in	1	2.5
Pressure regulator, 1–½ in	1 adjusts supply	0
Cross, 1–½ in	1	9
Elbow, 45 degrees	2	6
Swage, 1–½ × 3 in	1	1.5
Total equivalent feet of pipe		497

Figure 11.2 shows the unit configuration with the motor mounted over the pump and a vertical belt drive.

The belts were tightened, which seemed to solve the problem. During the final test, the flow was measured at 10 gpm under no load and 8 gpm when pressured to 6,000 psi; however, 8.8 gpm was required at 6,000 psi according to the contract. The belt/drive ratio was reviewed, and the pump efficiency was analyzed. With the plunger size being 25 mm, the flow should have been 0.0232 gpr. The motor speed was measured at 1,465 rpm, and the belt/drive ratio was 3.59:1.0 for 407 rpm to the pump-input shaft. Even though 407 rpm × 0.0232 gpr = 9.48 gpm, the pump discharge only measured 8 gpm at 6,000 psi, which is about 84% efficiency. This triplex, 60-mm stroke pump should have been 90% efficient or better, so the first thoughts were to check the flow meter at zero pressure and 6,000 psi by using a calibrated container. Other actions were discussed that included lapping the valves or replacing them, checking for leaks in the pump packing, considering an increase in the speed of the pump, or increasing the plunger size. Before any of these things were done, the belts were checked again, and the proper tensioning was

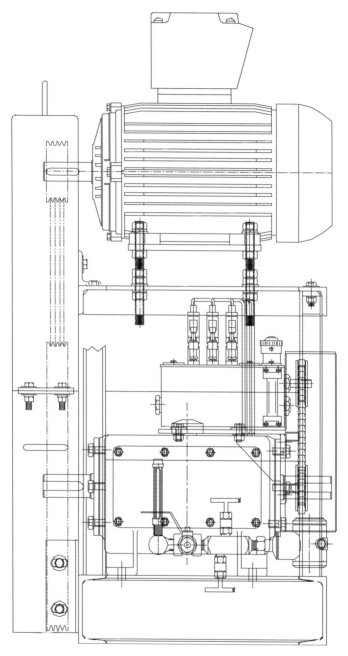

Figure 11.2. Pump unit with electric motor mounted over the pump

Tension Measurement By Deflection

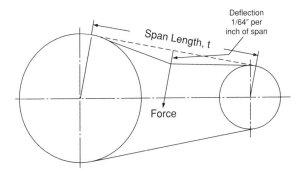

Figure 11.3. Belt-tensioning guide diagram

measured using a gauge. Another method for checking belt tension based on deflection is shown in Fig. 11.3. The pump was retested and delivered 8.7 gpm at 6,000 psi, which was acceptable to the customer.

11.8 Cockpit Problems

Customers who are new to high-pressure pump operation have very interesting comments and questions. For example, a pump operator who is having obvious problems with a system has been said to be having "cockpit problems." This comment suggests that an operator can have years of experience and still have limited knowledge of how a positive-pressure, reciprocating pump works. There are also times that special problems baffle the most experienced users, and these can only be solved by the process of elimination.

11.8.1 Not Enough Pressure

A high-pressure pump unit with a hose, a gun, and nozzles for water jetting was shipped to a customer. He connected the accessories, supplied water to the pump, and started the engine to see how the unit performed. The customer believed that the pump would only reach 70 psi, when it was suppose to operate at 10,000 psi. It took days of making phone calls, trying different nozzles, checking belts, checking pump valves, and checking packing until it finally was discovered that he was reading the wrong scale on the gauge. He was reading mega-pascals instead of pounds per square inch on the discharge pressure gauge, and the high-pressure pump unit had been running a steady 10,000-psi all along.

11.8.2   Bad Foot Gun

Another customer having cockpit trouble called a pump supplier to complain about the new foot-operated gun he had purchased from that company. During his phone call, this customer explained that his diesel engine would bog down every time he engaged the foot gun. After the company representative had him recheck the foot gun, the connections, and the type of work he was doing, the supplier realized that the customer had attached a small rigid lance to the foot gun and that the pressure drop through this lance was about 9,000 psi for a 10,000 psi pump. The pressure drop in a water-jetting setup of this kind can be checked by removing the tube nozzle at the end of the rigid lance or flexible lance then checking the gauge pressure while running the pump at normal speed. The gauge pressure is the same as the pressure drop in the system. The customer was then made aware that the foot gun was not defective.

11.8.3   More Cockpit Problems

Sometimes the obvious is not so obvious to the new operator. It is advisable to do a start-up test at the final destination of a pump unit, but this is not always possible. A new operator called a pump supplier to report that he had started the engine on his new pump unit, engaged the clutch, and increased the engine speed, but the gauge read zero. When asked if he had connected the gun and hose, he said that he had; however, when asked if he had triggered the gun, he answered, "No. Was I supposed to?" This step is not obvious to some users because some high-pressure pumps have a back-pressure regulator or unloader that maintains pressure in the system. A water-jetting system with an unloader would normally use shut-off-style guns. This customer's system included a dump-style gun, so the representative explained to him that high-pressure, positive displace-ment pumps produce flow and will only build up pressure when a valve, orifice, or other back-pressure device restricts the flow. With his cockpit problem solved, the customer could now fly.

11.9   Troubleshooting Guide

A chart of troubleshooting tips is usually included in a pump manual. The following shows an example of a chart that might be useful for an operator or mechanic.

11.9.1   Pump Manual Troubleshooting Tips

11.9.1.1   No Flow from Pump

- Tank is empty.
- Inlet valve is closed.
- Inlet strainer is clogged with debris.
- Crankshaft is not turning.

11.9.1.2   Insufficient Pressure from Pump ONLY

- Pump speed is too slow.
- Relief valve is improperly adjusted and bypassing fluid.
- Oversize or worn nozzle exists on the equipment.
- Pump valves are worn.
- There is excessive leakage from pump seals.

11.9.1.3   Insufficient Flow from Pump ONLY

- Pump speed is too slow.
- Relief valve is improperly adjusted and bypassing fluid.
- Pump valves are worn.
- There is excessive leakage from pump seals.

11.9.1.4   Insufficient Flow OR Pressure AND Rough Operation

- Valve problem:
  - Pump valve is stuck in open or closed position.
  - Valve assembly is damaged or unseated.
  - Valve seat is washed out.
  - All pump cylinders are not primed.
  - Inlet strainer is clogged with debris.

- Excessive gas in liquid:
  - Air leaks exist in suction line or fittings.
  - High spots in suction line are allowing formation of gas pockets.
  - There is a vortex in the tank near the inlet pipe opening.
- Cavitating pump:
  - NPSHa (tank head or charge pressure) is insufficient.
  - Fluid viscosity is too high.
  - Inlet line is too long and/or too small diameter.

11.9.1.5   Pump Running Roughly, Knocking, or Vibrating ONLY

- Plunger assembly is loose.
- Valve assembly is damaged or unseated.
- Pump is cavitating due to:
  - Insufficient NPSHa (tank head or charge pressure);
  - High fluid viscosity ;
  - Long length or small diameter of inlet line;
  - Worn or damaged power frame components; or
  - Air across worn piston cups.

11.9.1.6   Suction Pressure Fluctuating Rapidly

- Pump is cavitating.

11.9.1.7   Fluid Leakage from Pump

- Plunger packing is wearing and about to fail.
- Fluid cylinder bolts are not properly tightened.
- Fluid cylinder O-rings (or gaskets) are damaged.
- Piston assembly O-rings are damaged.

11.9.1.8   Short Plunger Packing Life

- There is a high abrasive particle content in fluid.
- Wrong style or type of plunger packing is being used for service.
- Plunger is damaged.
- Pump is cavitating (cylinders may run hot).
- Stuffing box assembly or seal is damaged.

- Poor quality water is being used.
- Pump is allowed to run dry for extended periods of time.

11.9.1.9 Short Valve Life

- There is a high abrasive particle content in fluid.
- Valve assemblies were only partially rebuilt during previous service.
- Valve assemblies were damaged due to improper installation techniques.
- Poor quality water was used.
- The pump is cavitating.

11.9.1.10 Cracked Fluid Cylinder

- Discharge pressure was too high.
- Pump was exposed to freezing conditions without properly draining.
- Hydraulic shock resulted from cavitation or entrained air.
- Discharge valve is stuck shut.

11.9.1.11 Crankshaft Jerking or Starting and Stopping Rotation

- V-belts are loose and slipping (if equipped).
- Hydraulic system relief valve is chattering (if equipped).
- Pump was operated at excessively high discharge pressure.
- Discharge line is blocked or partially obstructed.

11.9.1.12 Power End Overheating (in Excess of 180° F)

- Discharge pressure is too high.
- Low oil level exists.
- Improper oil viscosity exists.
- The power-end oil is contaminated.
- Pump speed is too fast.
- Pump is running backwards.
- Couplings are misaligned.
- V-belt drive tension is too tight.
- Pump is located too close to heat source.
- There are worn or damaged power frame bearings.

11.9.1.13   Broken Crankshaft or Connecting Rod

- Pump was exposed to freezing conditions without proper draining.
- Discharge pressure was too high.
- Suction pressure was too high.
- Hydraulic shock occurred due to cavitation.
- There is a material or manufacturing defect.

11.9.1.14   Broken Fluid-End Bolts

- Bolt or nut was not properly torqued.
- Discharge pressure was too high.
- There are excessive piping loads on fluid end.

11.9.1.15   Contaminated Power-End Oil

- Pump has been operated with failed plunger packing for extended periods of time.
- A high-pressure wash wand was used to clean near breather or oil seal areas.
- Deflector shields are missing or damaged.
- Crosshead extension oil seals are damaged or improperly installed.

11.10   Test Loops

Testing and troubleshooting are two areas that often fall back on engineering. Test loops are a good way to prove a principle or test a system for endurance; testing will prevent the agony of troubleshooting in the field after delivery of the equipment.

11.10.1   Methanol Test Loop

A vertical 60-hp pump unit was part of a methanol test loop in Louisiana when a customer required a 600-hour test of the pump under load. Pressure, flow, temperature, energy used, packing life, and performance were recorded during the testing period. The high-pressure pump skid, shown in Fig. 11.4, was connected to the test skid, shown in Fig. 11.5. The test skid consisted of a product tank, pressure recorder, gauges, filter, supply pump, and heat exchanger. The heat exchanger

Figure 11.4. Vertical triplex pump in methanol test loop

was required to keep the methanol at a normal temperature during the long tests. There was a successful conclusion to the testing that provided data for future designs and improvements to the high-pressure pump.

Figure 11.5. Methanol test system with supply tank and instrumentation

11.10.2　Packing Test Loop

Another example of a test facility, as shown in Fig. 11.6, was used to prove plunger-packing concepts. The three-barrel design of the pump allowed operation to a 15,000-psi working pressure and allowed the testing of more than one type of packing during the same test run. One stuffing box could have a standard packing design, while another box could have a new packing concept that needed to be tested. This configuration was used to test the *Hydro-Balanced Packing System* during the preliminary stages of development, as discussed in Chapter 9; the system was then used by a packing manufacturer in Houston, TX, for 2 years before it was moved to Lafayette to test a new packing system for methanol pumping.

11.10.3　Shop Testing

It is more common for a pump or pump unit to be shop tested after assembly by the manufacturer. A test report is usually completed for the customer and the job file. It has been the custom of some pump unit

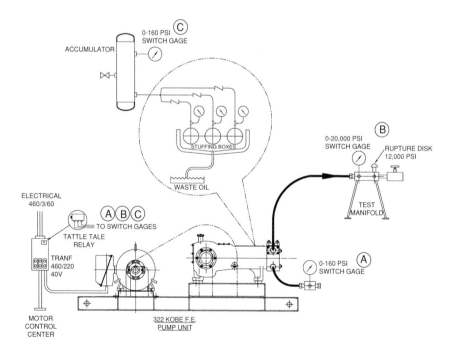

Figure 11.6. Methanol test loop for a horizontal triplex pump

suppliers that the test be limited to only a few minutes or last until the system stabilizes and it seems to be running well because it costs more to run long, comprehensive tests. It might make some sense to run short tests on an assembled unit when a pump and motor or engine has already been tested, but sometimes things happen in that third hour of continuous running that did not happen in the first 30 min. The problems may appear as leaks or noise, or they may appear as heat or vibration that is apparent during a longer test; these problems can be corrected before the equipment is shipped. A test report form can be simple or intricate according to the facility capabilities, but it should include a few basics. Items such as date, location, pump model, serial number, plunger size, stroke, pressure, flow, and speed should all be included on the test form. Figure 11.7 shows data on a test form that is set up as an Excel sheet; the data are placed into the matrix of test results, and the flow rate against the discharge pressure can be plotted. The Hydraulic Institute standards cover pump testing and discuss witnessing, instrumentation, procedure, and records, which can be further reviewed on their Web site.

11.11 The Correct Pump

Selecting the right high-pressure, positive displacement pump can be fairly straightforward, and the manufacturer or supplier can suggest a model that best meets the user's requirements. A step-by-step approach would include the information in the following subsections.

11.11.1 Horsepower Calculations

To calculate the horsepower range of the pump to be selected, one of these formulas based on the desired performance can be used:

Hydraulic horsepower = gallons per minute × pounds per square inch × 0.00058.

Hydraulic horsepower = barrels per day × pounds per square inch × 0.000017.

Hydraulic horsepower = gallons per minute × pounds per square inch/1714.

To approximate the mechanical and volumetric inefficiencies, the horsepower can be divided by 0.85 to determine what horsepower

Weatherford

Completion and Production Systems

TEST REPORT

TEST DATE :
LOCATION :   Houston
PUMP MODEL :   2100 R
PUMP TAG NO.:   NOTES
MOTOR TAG NO.:   * Acceptable operating range 8-10 GPM
STROKE LENGTH (MM) :   100   (4")   ** Acceptable Temperature range < 180°F
PLUNGER SIZE (MM) :   20   (.78") *** Acceptable Temperature range < 160°F
VALVE TYPE :   Disc   **** Acceptable Motor HP level < 100
FLOW TARGET (GPM) :   9   ***** Due to motor power limitations, Test
PRESSURE TARGET (PSI) : 15,000   pressure will be limited to 15000 PSI
GPR :   0.0250
SUCTION PRESSURE: 40.0000

TIME : (P.M.)	9:15	9:25	10:45	11:00	11:15
***** PRESSURE (PSI) :	500	5000	10000	15000	22,500
* FLOW RATE (GPM) :	8.2	7.57	7.46	6.94	6.24
** FLUID TEMP. (F) :	71.1	82.9	95.3	108.3	120.3
*** POWER END TEMP (F):	70				
INPUT SPEED :	1800				
PUMP SPEED (RPM) :	343	343	343	343	341
THEORETCAL FLOW :	8.58	8.58	8.57	8.57	8.52
VOLUMETRIC EFF. :	96%	88%	87%	80%	73%
**** POWER (HP) :	2.4	22.0	43.5	60.7	81.9
MOTOR AMPS :	48	58	68	84	118

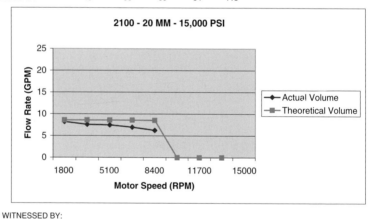

WITNESSED BY: _____

Figure 11.7. Sample of test report on a triplex horizontal pump

range to consider for the driving engine or electric motor. The resulting horsepower can be rounded up to match the next standard-size engine or motor. In addition to the horsepower capability of the high-pressure pump, the fluid that it is to handle is important. Oil-field pump fluid ends may be called *carbon steel*, referring to cast or forged steel blocks and are also available in aluminum bronze, 316 stainless steel, ductile iron, 17-4 stainless steel, Carpenter 20, and even titanium. See Appendix A for an example of a compatibility chart showing materials that can be used with various fluids to be pumped.

11.11.2 Suction Pulse Dampener

A suction dampener for the pump size selected should be designed to remove harmful surges in the high-pressure supply line. It increases pump efficiency and a longer life for the fluid-end components. By working with a pulsation dampener supplier or manufacturer, users can select the right type and proper size of the suction dampener for their project.

11.11.3 Net-Positive Suction Head

The application for the high-pressure pumps and the type of system in which it is to be used are very important considerations for proper performance. The suction pressure must allow the plunger pump to fill on each stroke, so certain characteristics need to be considered. Net-positive inlet pressure (NPIP) is measured on the absolute pressure scale. The absolute pressure is the sum of local barometric pressure plus the gauge pressure therefore, the minimum NPIP is the inlet pressure, which will allow a pump to operate with acceptable performance and good volumetric efficiency. It is affected by pump speed, plunger diameter, valve type, valve-through area, valve spring load, temperature, and vapor pressure of the liquid being pumped. One pump company suggests that 15 psig above the vapor pressure is required as a conservative measure for any plunger pump while running. Any NPIP above the minimum amount will result in improved fluid-end performance and can help with any unpredicted changes in operating conditions.

11.11.4 Fluid Supply

The supply tank elevation should be as high as possible above the suction when an open vessel is used to feed the positive displacement pumps. Pipe friction losses should be considered if the tank is located a long distance away from the pump or if there are fittings and valves in the supply line. A short inlet line is usually desired when using an atmospheric supply tank. Other methods of feeding the high-pressure pump may include a charge pump and/or a suction-side pressurized tank that should be sized to provide the NPIP needed. The liquid level above the pump inlet can be used to determine inlet pressure. Pounds per square inch = Inlet pressure = psi = feet of head × 0.433, and the atmospheric pressure can be calculated by pounds per square inch absolute = local uncorrected barometric reading (inches of mercury × 0.4912).

Atmospheric pressure is the uncorrected barometric pressure at the altitude of the pump in pounds per square inch absolute. Barometric pressure reported by the weather bureau is normally corrected to sea-level readings, and the uncorrected reading is to be used to determine the NPIP.

Gas blanket pressure on a supply tank adds additional positive pressure to the inlet of the pump. It also helps to degassify the liquid being pump. Some water-blasting pump unit manufacturers use a pressurized suction tank (with or without a bladder) to act as a suction pulsation dampener and stabilizer in the suction line to their pumps. A pressure gauge is installed in the tank to read the inlet pressure, which includes the gas blanket, but it seems like a mistake to put a relief valve at the top of such a tank that could exhaust the air above the water. The relief valve can be placed in the line connected to the tank to protect it from overpressuring.

11.12   Other Factors

Vapor pressure is the absolute pressure at which a liquid will change to vapor (boil) at a given temperature. Vapor pressure is important when pumping hydrocarbons such as propane, crude oil, butane, LNG, nitrogen, methanol, or water at elevated temperatures. Appendix B shows the pressure/temperature relationship of fluids.

11.12.1   Eliminating Friction

Friction in piping that supplies the fluid to the high-pressure pump is often the cause of pump problems. The inlet line should be as short as practical with a minimum of bends, valves, and fittings. When the same tank supplies more than one pump, it is best to run individual inlet lines to each pump. If bends and elbows are required in the system, they should be long radius not sharp bends to cut down on friction and turbulence.

11.12.2 Acceleration Head

Acceleration pressure or head is the inertia effect of the fluid mass in the inlet line. It is a function of the volume of the fluid in the line, the pump speed, number of plungers, and pump displacement. The pressure required to accelerate cold water without cavitation can be calculated with the equation:

$Pac = KNLQ$
= acceleration pressure in pounds per square inch absolute

In this equation, K is a constant for a triplex single-acting pump of 0.000259, N is the pump speed in revolutions per minute, L is the inlet line length in feet, Q is the pump discharge in gallons per minute.

The diameter of the inlet line is an important factor in the formula, but it has been shown that a suction stabilizer located near the high-pressure pump can greatly improve the acceleration pressure requirement.

11.12.3 The NPIP

The NPIP is the absolute pressure plus the vapor pressure required to fill the pump cylinders so that no more than a 3% drop in the pump capacity occurs during the run test. The NPIP available is the pressure in pounds per square inch absolute of liquid at the pump inlet. The minimum height in feet above the pump inlet can be determined by the equation:

$H(feet) = NPIP \ required + Pv + Pf - Pa) \times 2.3$

In this equation, Pa is the atmospheric pressure in pounds per square inch absolute due to altitude, Pv represents the vapor pressure in pounds per square inch absolute of the liquid being pumped, and Pf refers to the pressure drop in pounds per square inch caused by friction in piping and fittings. In addition, H is the height of liquid above the pump measure in feet, P1 represents the pressure in pounds per square inch gauge at the pump inlet, and P2 represents the pressure in pounds per square inch gauge at the tank outlet.

$$P_2(psig) = H \times .443$$
$$P_1(psig) = P_2 - P_7$$
$$or$$
$$P_1(psig) = P_1 + P_a$$

It is a good practice to add 5 to 10 ft of head to the minimum NPIP, and 15 psia has been suggested for a required NPIP on larger positive displacement pumps. A centrifugal charge pump can increase inlet pressure when NPIP is insufficient. It has been said that the total horsepower requirements will be the same because the high-pressure pump will require less power input when a charge pump is used. The high-pressure pumps and UHP pumps, along with pressurized inlet tanks and/or pulsation dampeners, usually have a positive feed of 20–75 psi or more.

11.12.4 High Inlet Pressure

When the inlet pressure is too high, it increases the bearing loads and requires a pump resizing or smaller plungers if the inlet pressure is more than 5% of the discharge pressure. The equation for a triplex single-acting pump is as follows:

$$Px = Pd + 1/2Pi.$$

In this equation, Px represents the equivalent discharge pressure of the pump, Pd is the actual discharge pressure, and Pi is the actual inlet pressure.

The equation for a quintuplex pump, which is listed here, uses the same values as noted in the previous equation:

$$Px = Pd + 2/3Pi.$$

The work performed by the pump is calculated in the usual way by using actual discharge pressure less the inlet pressure or Delta P. For example, high inlet pressure was considered when a customer wanted a 10,000-psi discharge pressure at 198 gpm and an inlet pressure of 6,000 psi. The following equation was used for this calculation:

$$Px = Pd + 1/2Pi.$$

Hence, for a triplex pump, the artificial pressure is $Px = 10,000 + 3,000 = 13,000$ psi (the pressure to be used to size the pump). A flow of 198 gpm at 13,000 results in 1500 hp, and the rod load for 2-in plungers is 40,840 lb. If the customer's requirements were 10,000 psi at 198 gpm, with a normal inlet pressure range, the horsepower would be calculated as 1155 hp, and the rod load for a 2-in plunger would be 31,415 lb. It is predicted that the pump would fail early if the excessive inlet pressure were not considered.

11.13   Important Factors to Remember

It cannot be overemphasized that the performance of a high-pressure pump can be affected by several factors. The inlet line supplying a pump should be as short and direct as possible. The diameter of the inlet line should be one size larger than the suction manifold connection, especially when the suction pressure is low. Inlet piping should be free of restrictions; fittings should be long radius, and full, open-type valves should be used. The discharge piping should be the same size or larger than the manifold connection. It should allow full discharge flow and permit no more than a 12-ft/s velocity. Its fittings should be long radius with no more friction than a 45-degree elbow.

In addition, bends should not be installed adjacent to the pump discharge connection; all pipes, fittings, hose, and devices should have a working pressure at least equal to the maximum pressure rating of the pump. Bypass devices can permit starting the high-pressure pump under no load or adjusting the pressure in the system. The discharge from the by-pass device should not be piped back to the inlet of the pump, but it can be plumbed to an atmospheric supply tank or drain. At least one relief valve or relief device should be mounted on the discharge line as close as possible to the pump discharge connection. The pressure-relief device on the fluid end of the high-pressure pump or in the discharge plumbing should be set no more than 1.25 times the rated working pressure. The discharge from the pressure relief device can be plumbed back to an atmospheric supply tank or to an open drain for maximum safety.

11.14   Lubrication

Maintenance of a high-pressure pump includes using the proper lubricant and the correct replacement interval. The chart in Fig. 11.8 has more information than found in the usual pump manual. Generally,

MAINTENANCE & LUBRICATION RECOMMENDATIONS

DESCRIPTION: The oil is suitable for application in the temperature range of 10-deg. F minimum ambient temperature and 225-deg. F maximum bulk operating oil temperature. However, normal operating temperature for Triplex Pump Crankcase Oil is from 160-deg. F to a maximum of 180-deg. F.

TYPICAL CHARACTERISTICS: The physical and chemical characteristics of this oil are shown below. These Oils are premium quality, lead free, heavy-duty industrial gear lubricants. Additives are included to provide extreme pressure and anti-wear characteristics, rust and corrosion protection, increased oxidation stability, and improved resistance to foaming and excellent high load performance.

APPROVED LUBRICANTS

CHARACTERISTICS	MOBIL GEAR OIL 630	EXXON SPARTON EP 220	TEXACO MEROPA 220	SHELL OMAL A 220	CHEVERON NL - GEAR **LUBE** 220
GRAVITY - °F API	26.5	24.1	26.2	25-27	26.0
SPECIFIC GRAVITY	0.896	0.894	---	---	---
POUR POINT - °F(°C)	0 (-18)	0	0	0	0
FLASH POINT - °F(°C)	425 (218)	415	450	425	440
VISCOSITY	---	---	---	---	---
SUS at 100 °F (37.8 °C)	1045/1165	1166	1080	950/1050	1050
SUS at 100 °F (37.8 °C)	92	88.2	90	89/95	104
CST at 40 °C	198/220	220	220	205/227	198/220
CST at 100 °C	18.0	17.0	17.5	17.4/19.2	18.0
VISCOSITY INDEX	95	90	92	98	95- MIN.
COLOR ASTM	7.0	7.0	---	5.5/ 7.0	8+
CHANNEL POINT- of (°C)	-25 (-32)	25	---	---	---
RUST - ASTM D 665	PASS	PASS	PASS	PASS	PASS
TIMKIN OK LOAD – LB. , ASTM D 2782	70	60	60	60	60
US STEEL EXTRA- DUTY GEAR OIL NO. 222	PASS	PASS	PASS	PASS	PASS
US STEEL EXTRA- DUTY GEAR OIL NO. 224	PASS	PASS	PASS	PASS	PASS
EZG TEST- NO. STAGES PASSED	13	11	11	11	11
ISO / ASTM VISCOSITY GRADE (VG) °C	220	220	---	---	---
AGMA LUBRICANT NO.	5 EP	5 EP	5 EP	5 EP	5 EP

Figure 11.8. Sample lubrication chart for Kobe vertical pumps

the pump manufacturer will specify the grade, weight, and maybe even the brand name that is recommended for the pump. The physical and chemical characteristics of the lubricants are shown on the chart.

The oils shown in Fig. 11.8 are suitable for application in the temperature range of 10° F minimum ambient temperature to 225° F maximum bulk-operating oil temperature. However, normal operating temperature for the triplex pump crankcase oil is from 160° F to 180° F. These oils are premium quality, lead-free, heavy-duty industrial gear lubricants with additives to provide extreme pressure and antiwear characteristics, rust and corrosion protection, increased oxidation stability, improved resistance to foaming, and high-load performance. One pump manufacturer recommends grade 4 turbine oil in the 150–700 viscosity range, and one of the company's main OEM accounts uses *Texaco 80W90* gear oil. To explain, grade 4 turbine oil with 150 viscosity is in the range of 90-weight gear oil, which is equivalent to 40-weight motor oil. Figure 11.9 is a comparative viscosity classification chart that can clarify discussions concerning the viscosity of motor oil versus gear oil.

11.14.1 Plunger-Packing Lubrication

Some high-pressure pumps have a forced-fed oiler installed on the power frame and driven by the pump crankshaft, as shown in Fig. 11.10. A packing lubricator can be driven by the high-pressure pump crankshaft or independently driven by another source. In Fig. 11.11, an extra 5-gallon oil capacity has been added to increase the run time of the pump without adding oil to the mechanical lubricator.

The reservoir on the lubricator has a capacity of 4 pints of oil, and there are approximately 14,500 drops in a pint, so the reservoir contains about 58,000 drops. Setting the rate at 6 drops per minute, the reservoir should supply oil to the plunger packing for a period of 9600 min (161 hours or 7 days) of operation. The 5-gallon supply tank plumbed to a flow switch adds oil automatically to the 4-pint reservoir. The mechanical lubricator can handle almost any oil, including the same oil used in the high-pressure pump crankcase; however, for optimum packing life, one of the following is recommended. Note that these lubricants can be used in both winter and summer months with an ambient temperature of 10° to 225° F.

- **Nonsynthetic plunger lubrication:** *Rock Drill Oil #46* with operating temperature range of −20° to 210° F.
- **Synthetic plunger lubrication:** *Pinnacle Lubricant #46* with a pour point of −50° F and an operating temperature range of 210° F.

Maintenance on a high-pressure pump includes repairing anything that wears or fails in service. It also includes changing the crankcase oil

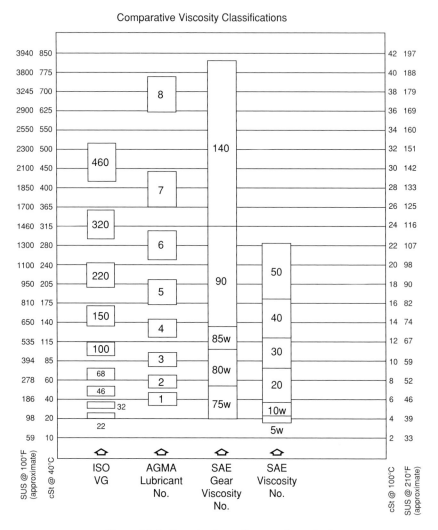

Figure 11.9. Sample of an oil comparison chart

at the recommended intervals or changing the oil when it is milky, which indicates water in the crankcase.

11.15 Pump Storage Considerations

After high-pressure pumps are built and tested, it is sometimes necessary to store them for a period of time. Customers may require a procedure for the preservation of their high-pressure pumps until

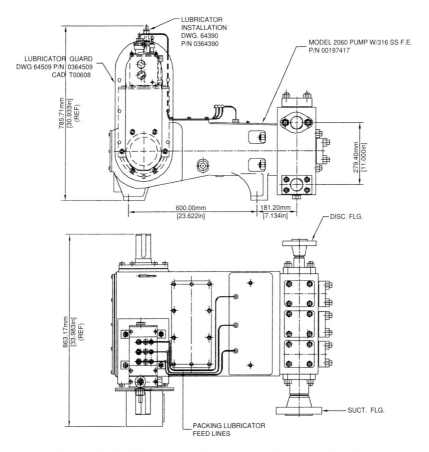

Figure 11.10. Triplex pump with a mechanical packing lubricator

they are to be used. It is not uncommon for pumps to be stored in less than ideal conditions for 1 year or more. The following paragraphs provide information for preventing expensive damage to a pump that must be stored for a long period of time.

11.15.1 Storage Definition

- *Long-term storage* is defined as any continuous period of non-use for reciprocating pumps that is greater than 3 months. The system should be placed in an appropriate storage environment. This time period should include time spent for shipping and warehousing and should also include the time for pump/system shutdown.

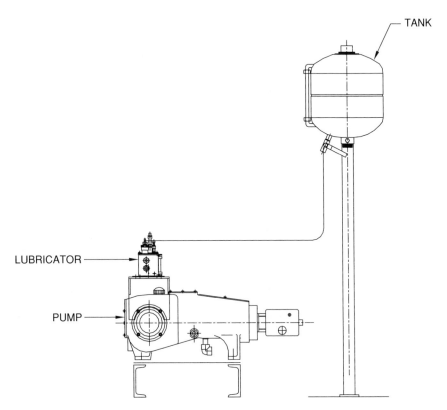

Figure 11.11. Mechanical packing lubricator with extra oil capacity

- This storage procedure applies only to reciprocating pumps. Long-term storage of other unit components (e.g., electric motors, gear reducers, engines, etc.) must follow the specific manufacturer's recommendations.

11.15.2   Storage Environment

The storage/shipping conditions must at least protect the pump from sun, rain, wind, sand, and other debris. In addition, the pump should be protected from large temperature changes and high humidity levels.

11.15.3   Preparation for Long-Term Storage

It is important to clean the internal and external surfaces of the pump. In addition, drain all oil and remove all sludge and dirt. Remove and clean the oil-level gauge, oil breather/filler cap, and any other accessories. Plug all holes with pipe plugs or an appropriate air-tight seal, including suction and discharge connections. Fill the pump's power end to the normal operating level with the preservative and lubricating oil mixture prescribed in section 11.5.3.4 under the topic of preservatives and lubricants.

11.15.3.1 Steps to Take for Proper Long-Term Storage Lubrication

- With the oil mixture in the pump, rotate the pump crankshaft through 10 to 15 complete revolutions. This will coat the crankshaft and bearings.
- Leaving the oil mixture in the pump, inspect the gaskets, plugs, and seals for leakage. If leaks are observed, follow the instructions in the operation and maintenance manual to fix the leaks.
- Make certain that the pump's outer surfaces are clean. If painting is required, mask off the crankshaft, plungers, and any exposed seals or bearing surfaces. Paint as required.
- Apply a thin layer of the oil mixture to the exposed surfaces of the crankshaft and plungers. (Note that it is important to coat the nondrive shaft for pumps with double-extended shafts.) Wrap the crankshaft with a wax tape.
- Carefully wrap and place all components removed from the pump into protective packaging.

11.15.3.2 Periodic Inspection of Pumps in Long-Term Storage

Inspect all pumps in long-term storage on a monthly basis. During each inspection, perform the following tasks:

- Look for any sign of fluid leakage.
- Look for any sign of corrosion.
- Rotate the pump crankshaft 10 to 15 complete rotations.

Note that the preservative and lubricating oil mixture must be drained from the pump and replaced at least once every year.

11.15.3.3 Start-Up of Pumps after Long-Term Storage

- Prior to pump start-up, drain all of the preservation and lubricating oil mixture from the power end.

- Inspect all seals and all components removed from the pump. Replace as required.
- Remove the pipe plugs, and suction and discharge port seals. Re-install the drain plug, oil level gauge, oil breather/filler cap, and any other components removed for storage.
- Fill the pump power end with the proper type of lubricant and to the indicated level as listed in the manufacturer's manual.

Once these steps are completed, follow the pump start-up procedures outlined in the manufacturer's manual.

11.15.3.4 Preservatives and Lubricants

One of the preservatives that could be used during long-term storage of a positive displacement pump is manufactured by Daubert Chemical Company and is called *NOX Rust VCI 105*. This product or an equivalent product should be added to the pump crankcase normal lubricant in the amount of 1–2% by volume. Other lubrication products for the pump crankcase, such as from Royal Purple, have been reported to improve pump performance and should be considered for the possibility of increasing bearing life, lowering temperature, and extending oil-drain intervals.

11.16   Preventive Maintenance

Preventive maintenance can decrease the amount of trouble shooting for a high-pressure pump. A regular program includes daily or monthly maintenance for some high-pressure pump systems. Steps involved in such a program are listed here:

- A daily check of the crankcase lubricating oil is recommended so that the proper level can be maintained. The condition of the oil should be observed, and the proper oil as recommended by the manufacturer should be used. Pumps with stuffing box lubrication may have grease fittings or a forced-fed lubricator that should be serviced or checked daily. Packing nuts on pumps with spring-loaded packing should be kept tight or adjusted to stop excessive leaking on pumps with non–spring-loaded packing.
- Monthly maintenance includes changing the crankcase oil at the intervals recommended by the manufacturer or when foreign

matter enters the crankcase. The crankcase air filter screen should be kept clean by washing it in kerosene, or the filter element should be replaced when dirty. All studs, nuts, and cap screws should be tightened to their proper torque. Gaskets, seals, and connections should be checked for leaks and serviced when necessary.

- General maintenance and overhauls before serious troubles occur will reduce operating costs. At any sign of a pump problem, the source should be isolated. Valves and springs can be checked and replaced if worn, bent, grooved, or broken. Connecting rod wrist pins and connecting rod shell bearings can be checked for excessive play so they can be replaced if necessary.

11.17   Packaging Problems

Many problems can arise as a result of a packager buying pumps based on flow and pressure (horsepower) but do not follow good system design rules for high-pressure pumps. For example, one packager of reverse osmosis equipment purchased two pumps, packaged them, and shipped them to Bora-Bora. Nearly 1 year later, the system was started up, and the pumps ran for 2–3 hours before the packing leaked and at least one adjusting nut came completely loose. With adjustable packing, the pump needed to be checked at start-up and after about every 4 hours of operation to keep the packing nuts tight. Locks were not supplied with these pumps, but it would have been helpful to keep the adjusting nuts in place.

Another packager purchased two pumps for methanol service up to 15,000 psi and completed the packaging with autoclave tubing and fittings. One problem originated from the 12 elbows that were used in the suction line and the discharge line. The expected flow was not achieved during the actual operation offshore, so the problems haunted the manufacturer and not the packager. Troubleshooting includes working with the customer or operator to solve his or her problems and get the high-pressure pump back into service.

Chapter 12

High-Pressure Pump Systems

12.0 Introduction

Systems that include a high-pressure pump have become an important part of product manufacturing, material processing, fluid conveying, and plant maintenance. At first, a pump system used for cleaning consisted of handheld wands, such as a system used in 1947 to clean a cooling tower with 400 psi at 10 gpm. The high-pressure water-blasting handgun has its origins in the 1950s and has been redesigned and improved ever since. In the 1960s, some steel mills used high-pressure pumps with a fixed nozzle system to descale billets and slabs of red-hot steel. Descaling forgings with 10,000-psi nozzles now can be accomplished with automated systems that include rotating jets, manipulating equipment, and water recycling.

The development of automated systems using high-pressure water jets may have started in the 1950s and 1960s and progressed rapidly in the 1970s and 1980s. Mechanical devices can handle larger horsepower pumps with high flows and/or high pressures, improve safety, increase speed, and add to the efficiency of the cleaning process. The ultra-high-pressure (UHP) intensifiers and pump manufacturers seem to have taken the reverse direction of development by starting with highly sophisticated automation and then developing portable equipment with handheld accessories.

Heavy industry tends to have specific applications that must be done on a regular basis and lend themselves to the development of automated high-pressure water-jetting systems. Chapter 6 discusses an automated water-jetting system that removes high explosives (HE) from munitions while capturing the water and debris. In industrial applications, systems can be permanently installed and may include electric-driven high-pressure pump units, powered movement of the cleaning nozzles, and the manipulation of the item being cleaned. In the case of a large but portable item to be cleaned, a cabinet or special room can be designed so the work moves through the cleaning chamber. Good examples are found in the automotive industry—auto bodies, paint carriers, and conveyors are cleaned in automated booths.

Another example of a handheld process that developed into an automated method of cleaning can be found in the petrochemical industry. In the 1970s, chemical plants and their contractors used handheld guns to clean chlorine cells, but in the 1980s, portable air-powered fixturing devices were available to rotate the nozzle while moving it vertically and horizontally. With the installation of an electric-powered high-pressure pump, the system could be installed with a remote operator-controlled cleaning station. In addition to the nozzle movements, the rotating lance could move into and out of the chlorine cell being cleaning, thus allowing an X, Y, and Z coverage.

12.1 Chlorine Cell Cleaner

By the 1990s, chemical plants were more familiar with the benefits of automating some of their cleaning jobs. In one case, a customer requested a system to clean large chlorine cells that are used for making the gas from salt brine. The customer had been using a high-pressure pump unit and a handheld gun capable of 10,000 psi at 10 gpm to remove the asbestos coating from the cell anodes and now wanted a system that would clean the cells while rotating them for access to all sides. The ideas that were discussed included:

- A cleaning lance that moves in and out and that rotates the high-pressure nozzle;
- An indexer that moves the lance up, down, and from left to right;
- A work table that supports and rotates the large chlorine cell for cleaning;
- A control console to operate all the automated functions of the process;
- A 300-hp electric-driven high-pressure pump unit for 10,000 psi operation; and
- A hydraulic power unit to operate the system movements.

The complete system designed with these parameters was installed in the customer's facility, which included a containment room for the cleaning process, a control room for the operator, and a pump room for the high-pressure pump unit capable of producing 10,000 psi at 42 gpm. The increased horsepower available for cleaning and the automated system improved the cleaning effectiveness. The customer's personnel were exposed to less danger because they were protected in a control room when operating the equipment. The *Chlorine Cell* rotation device is shown in Fig. 12.1, and the *System Control Console* during testing is shown in Fig. 12.2. The device called an indexer to move the nozzle for cleaning the chlorine cell is shown in Fig. 12.3, and the 300-hp unit is shown in Fig. 12.4.

12.2 Water Expeller Cleaning System

The water expeller cleaning system originates from another customer's request. This customer wanted to automate a process to clean rubber dewatering equipment. A piece of equipment called the *Continuous Expeller* is used to squeeze the water from rubber being made by this plant facility. The main problem was that the rubber extruded out of the narrow openings in the expeller during the dewatering process, so a regular cleaning was required to keep the equipment operating. The job was not pleasant using manually operated water-blasting equipment and handguns because of the heat and humidity in the expeller equipment room. To address these problems, the customer requested the following features. The automated equipment should:

- Be able to run continuously in a dirty, hot, and humid atmosphere;
- Have a cleaning cycle of about $\frac{1}{2}$ hour for each expeller;

Figure 12.1. Chlorine cell cleaner rotating device

Figure 12.2. Chlorine cell cleaner system console

Figure 12.3. Chlorine cell cleaner nozzle lance indexer

Figure 12.4. Chlorine cell cleaner 300 HP pump unit

- Include an electric-driven high-pressure pump unit with minimum flow; and
- Be removable for servicing each expeller.

The first phase of the proposal included a design concept for the cleaning mechanism, a recommended pump size, and a budget price for the system to clean four of the expellers. Figure 12.5 shows the concept proposed for the cleaning device to be pneumatically powered and to be small enough so that it would fit inside of each expeller cabinet.

There were four expellers, and each side needed to be cleaned; therefore, eight cleaning mechanisms were proposed that would automatically move from side to side while moving vertically to cover the entire side of the expeller. The air logic would allow for continual operation of any pair of the cleaning mechanisms selected. Running only two of the mechanisms at a time could use minimum water and minimum pump horsepower. The high-pressure pump that was first proposed was a 150-hp electric motor-driven unit for 10,000 psi at 20 gpm. Later tests and prototype hardware proved that 20,000 psi at less than 10 gpm would do the job. After reviewing the concept, it was decided that a programmable linear controller (PLC) along with the automated valves and air logic would be desirable to control the functions of the system, as shown in Fig. 12.6.

The proposed operating procedure included a selection for manual or automatic operation as indicated by the following list:

- Supply electrical power to the system (pump and PLC) following company procedure.
- Select manual or automatic operation of the system.
- On manual, operate the equipment as desired. This includes opening the water-supply valve for the high-pressure pump, selecting one of the four cleaning devices, starting the high-pressure pump unit motor, and operating the system valve for pressure.
- On automatic, the start cycle button sequences the water supply to the high-pressure pump, directs the high-pressure water to the first cleaning device, directs operating air to the first cleaning device, starts the high-pressure pump motor, pressurizes the system after a time delay, runs the system for a preset time, depressurizes the system, switches to the next cleaning device, and repeats the cycle for the next three devices. When the automatic cycle is completed, the system depressurizes, the device movement is stopped, the high-pressure pump motor is stopped, and the water is turned off.

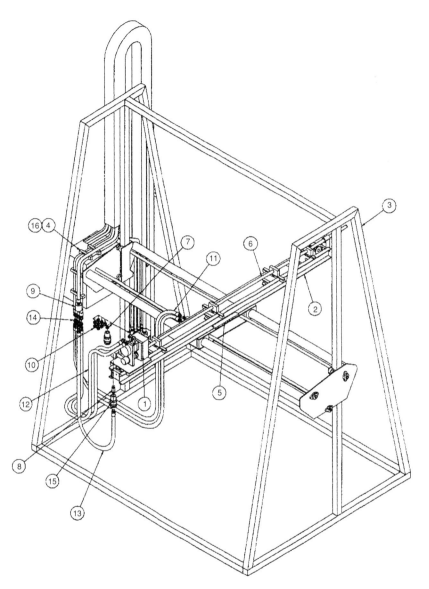

Figure 12.5. Water expeller cleaning device

The system is then in standby mode until the electrical power is locked out.

The final design for this system changed when the customer agreed to remove the cabinets over the expellers to allow better access. The

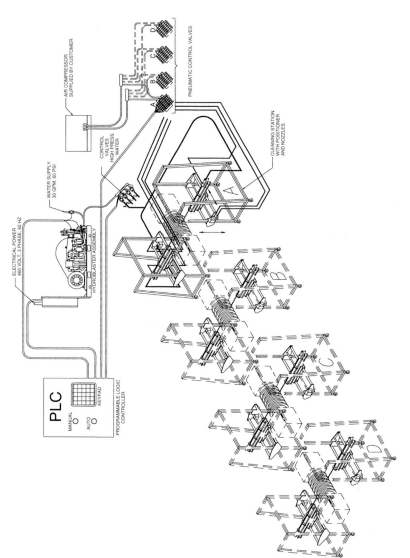

Figure 12.6. Water expeller cleaning system air logic

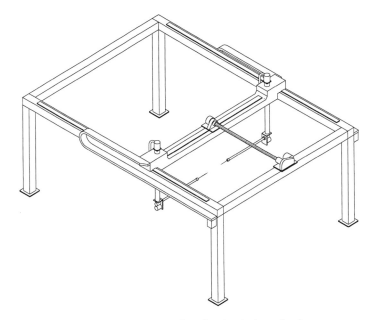

Figure 12.7. Water expeller cleaning test mechanism

mechanism shown in Fig. 12.7 was tested on the job site and was developed into the complete system.

12.3   Runway Cleaning Equipment

12.3.1   The Evolution of Runway Cleaners

One of the early runway-cleaning high-pressure pump systems was seen at Aqua-Dyne in 1977 when the company was working with a customer named Bob White. It seems that no one before White re- members runway cleaning using high-pressure water. He started by using a big rig, as shown in Fig. 12.8. The first pump units, with their large horsepower requirements, used a lot of water in the 5,000-psi range to clean rubber from the runways. An early unit owned by Bob White is shown in Fig. 12.9, and a newer version built by Aqua-Dyne is shown in Fig. 12.10.

This type of runway-cleaning equipment is expensive to buy and operate, but it also takes extra time to supply the large amount of water used during operation. Next came a unit that could reach a pressure of 10,000 psi, but it still used large quantities of water without

Figure 12.8. Runway cleaning equipment truck (*G. Pegoda*)

Figure 12.9. Runway cleaning operation (*G. Pegoda*)

Figure 12.10. Runway cleaning operation (*Courtesy of Aqua-Dyne*)

satisfactorily removing rubber and paint strips. This type of equipment progressed to using 400–500 hp and up to 15,000 psi at 50 gpm without the desired results. Then, UHP pumps using a pressure of 60,000 psi were tried using spray-nozzle configurations that tended to polish the runway surface and still not produce the proper cleaning.

Finally, the use of units that reached 200 hp with the right nozzle technology proved to be a better way to go. The latest hardware includes one or more rotating spray bars with a pressure to 40,000 psi to remove difficult deposits without damaging the runway. When cleaning, polishing, and thus possibly damaging, the surface is not desirable, but a good coefficient of friction for aircraft should be left. The nozzles used on the spray bars are smaller for good cleaning without cutting into the surface being cleaned. A modern version of the large *Airport Runway Cleaner* includes a large rotating cleaning head that is stored and can be operated at the rear of the vehicle. The unit can be operated from the cab of the truck with the electric controls, including the use of a closed-circuit television. The rotating cleaning head can also be used at the front of the truck, or a spray bar can be used instead, as shown in Fig. 12.10. The rotating head is shown in the rear operating position in Fig. 12.11. The right idea for modern times may be to reduce the horsepower, flow, and size of the equipment for a runway cleaning business, as shown in Fig. 12.12. This unit uses a spray bar for some of

Figure 12.11. Runway cleaning equipment truck (*Courtesy of Aqua-Dyne*)

the work, such as the washdown process, by mounting it to the side of the trailer. The rotating head is used for heavier cleaning, such as rubber and paint removal, as shown in Fig. 12.13.

Figure 12.12. Runway cleaning equipment trailer (*Courtesy of Aqua-Dyne*)

Figure 12.13. Runway cleaning lawnmower-type equipment (*Courtesy of Aqua-Dyne*)

In the early 1980s, White walked into Tritan and made a deal with the owner to build a runway cleaner under the condition that the owner would send someone to Saudi Arabia for start-up and training. Gilbert Pegoda was chosen to make the trip, and he met White in London in the spring of 1982, and they spent a couple of weeks together preparing for the equipment start-up. White explained how to clean a runway in technical terms even though it seemed to be a simple matter. According to White, he was the expert in runway cleaning, and Pegoda admits that what he explained did help on the job site. White only went with Pegoda as far as Remade, and Pegoda was left to run the system and train the local operators.

Most of the work for the project in Saudi Arabia was done at night because of the high temperatures. Normal daytime temperature in the desert was 115°–120° F, while the daytime temperature could be 130°–140° F on the runway. At night, the temperature fell to around 75° F. Because pictures for the figures had to be taken during the day, the figures show the water spray turning into steam. Pegoda oversaw

Figure 12.14. Runway cleaning in Saudi Arabia (*G. Pegoda*)

the removal of $\frac{1}{8}$ in or more of rubber buildup on the runways in Remade, Jetta (on the coast), and Mecca. The runways were extra large for heavy planes like the Boeing 747, perhaps because international travel is common in that part of the world. The equipment, as shown in Fig. 12.14, had two diesel engine-driven pump units that reached a pressure of 10,000 psi at 20 gpm and had a water tank mounted on a Peterbilt truck.

During the cleaning, it was determined that 5,000 psi would do the work, and the volume of water tended to wash the rubber to the side of the runway. The truck included a front-mounted spray bar with five fan-jet nozzles for a 24 inch path, as shown in Fig. 12.15.

At night, the lights aimed at the spray bar enabled the operator to see the path being cleaned. Wet rubber on a runway can be very dangerous, so sometimes the early morning dew precluded aircraft landings. The Saudi Arabians tested the runways for safety by using a new Saab automobile outfitted with a hydraulic apparatus including an aircraft tire on its back end. The car would be driven as fast as possible (maybe at 100 mph), the tire was lowered onto the runway, and then brakes were applied to the tire. By measuring the distance it took to stop the car, they would determine if the rubber buildup had become dangerous for landing aircrafts.

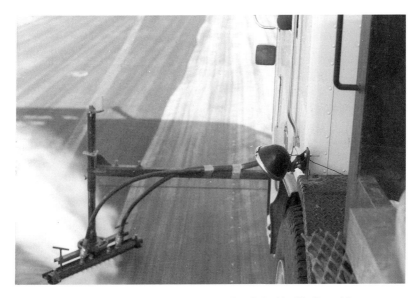

Figure 12.15. Rubber removal in Saudi Arabia (*G. Pegoda*)

12.3.2   Runway-Cleaning Equipment from Other Companies

Gardner-Denver (includes the companies of Partek/Butterworth, CRS, and Jetting Systems) builds custom packages, including runway-cleaning equipment. In the late 1970s, Partek built runway-cleaning equipment with two 150-hp pumps for use at 10,000 psi and overlapping fan-spray jets on a spray bar. Larger pumping systems evolved to 400–500 hp, and much thought was given to matching the pumps, water tanks, and horsepower for the jobs. Because the equipment is secured and the cleaning heads are mounted (not handheld), worker safety is not usually compromised.

Rampart Services (part of Flow International) is known as one of the first companies that suggested high-pressure specifications for runway cleaning. The company used intensifiers to remove rubber from runways and to remove paint strips in the 1980s. The larger units cleaned the runways very efficiently and rapidly so that the runways did not need to be closed for extended periods of time.

The industry (U.S. market) became very competitive in the 1980s, so to hold down cost, a contractor could purchase the pump unit and furnish his own truck, water tanks, and assembly labor to perform the job. In addition, the contractor had to have a vacuum truck or vacuum system to suck up the rubber because one of the federal agencies ruled that all the rubber had to be captured or removed from the cleaning operation. This process would help prevent ground pollution and negative environmental effects on animals and surrounding vegetation; however, this regulation made the operation more complicated and costly in the United States. After 20 years of selling the pump and components only, Gardner-Denver sold a truck-mounted unit in December 2001 for use in Egypt; the pump could operate at 30,000 psi and 12 gpm for airport runway-cleaning projects.

Reliable Pumps has been involved with runway-cleaning equipment since around 1985. Some of its first customers were involved in the runway-cleaning industry; many of its original customers have left the industry, however, because of heavy competition and high cost of investment. Reliable Pumps learned that runway-cleaning contractors use 10,000–20,000 psi and even higher pressures for rubber removal, paint removal, and cure removal. Several companies have been and are using high-pressure systems to perform these jobs.

In 1993, a company named J. M. Concrete was pouring concrete on runways in Albuquerque, NM, and was using a pump unit to deliver 12 gpm at 10,000 psi to remove the cure from the runways. When concrete is poured on a runway, the concrete cure needs to be removed so that paint will adhere to it.

Rubber removal contractors stopped using chemicals to remove runway rubber because the chemicals would dissolve the rubber, and it was a hazard because it was slick, so it became more acceptable for these companies to use straight water to remove the rubber.

Users of high-pressure water jets for removing rubber also include personnel who maintain Houston Intercontinental and heavy traffic areas, such as those in Florida. The work is hard because the cleaning may involve 24 hours of work per day for 1 week. These users have learned that a dependable pump is most important. The equipment may include a 6,000–8,000-gallon water tank on the truck and use the airport sweepers to collect the rubber—proper disposal of the water and excess rubber and debris has become important. When the operators run low on water, they must drive over to the source and then return to the work. Reliable Pumps has developed hardware for other markets because of their involvement with was originally developed for the runway-cleaning operations. The company offers a swivel that is used in the cleaning head (Hydro-Mowers). One or more cleaning heads may be located across the front of the truck. The cleaning rates have increased because of these developments, and the contractor gets more money, but it is a highly competitive market. Some of the contractors have two pump systems on their trucks just in case one pump needs maintenance and cannot operate; the contractors can perform the job with one system while working on the other system.

In terms of cleaning airport runways, companies find the most business on the East Coast and in the South. According to Russell Reed of Reliable Pumps, a company called Rampart out of Dallas and a company called Centerline are among Reliables customers. Centerline completed a job in Ft. Smith, AK; this area had a 12–15 mile run of highway concrete with cure that needed to be removed. A high-pressure pump reaching a pressure of 15,000 psi at 14–15 gpm was used to perform the task; the pump was attached to a truck that drove the distance of the centerline. A truck behind the first one painted the stripe as soon as the concrete was dry and put in a snowplow marker. Then a water-blasting gun was used on the return trip to clean for lane marking stripes.

There has been a great learning curve in this field, and users have learned from experiences and mistakes. Cleaning airport runways is not always a safe job—stories have been told of planes being accidentally directed to land on strips that were occupied by contractors and runway-cleaning equipment. The contractors had to move quickly to avoid being injured. Operators on the job also have to be alert and well-trained. If an operator does not know what he is doing or falls asleep on the job (as has been the case for some projects), the concrete can be damaged to a degree that it has to be repaired or repoured. Exposing the aggregate in the concrete may cause problems for landing and future cleanings. Contractors need to know the responsibilities of their job—for example, the task might not call for removing the paint strips, but if the contractor did so, a painting company's services would be retained—at the contractors expense. In addition, a contractor must be familiar with adjustments such as cleaning head height, speed, and intensity, which are very important for accomplishing a task properly and require a user's experience.

12.3.3   Water-Blasting Equipment and Processes

Sandblasting was once used for cleaning runways, but this technique is no longer in practice because of associated hazards with jet engines. However, it is important for water-blasting contractors to realize that even water jets can produce similar hazards after the water dries on the runway.

When performing a job using water-blasting equipment, the operator adjusts the spray bar (such as in the *Hydro-Mower*) by physically determining the proper height and judging the conditions. Someone else moves the truck to test the settings. Some of the rubber can be very thick, and its thickness and condition (e.g., grooves in runway) determines the measures for the optimum cleaning. A contractor, however, may not always realize how much time or material a project will require. For example, for a paint-removal project at the St. Louis airport, the contractor ended up filling four roll-off boxes with the amount of paint removed. This end result was not expected.

For runway cleaning, a 30,000-psi jet works fine, and sometimes a 20,000-psi jet does the job. Nozzles are important for each task, and Reliable recommends 12–15 quality nozzles (such as jeweled or tungsten carbide) for a self-rotating spray bar. The bounce back from the cleaning can damage a nozzle, but the higher quality nozzle lasts longer. A contractor will usually have a good stock of everything he might need, including different nozzles, because there is no time for securing things that are needed during the operation. Unusual things can happen on the job site, and the contractor understands that good equipment is required. The job might be paid according to a fixed amount of work or time or by the square foot, so speed is essential.

Reliable been selling the pump unit and the cleaning heads; the contractor has been running the hoses and controls. In terms of upkeep, contractors may drive their truck to a company, such as Reliable, for installation of certain units or even for maintenance.

A lesson that was learned the hard way involves the movement of the water in 6,000–8,000-gallon tanks used during a cleaning operation. When the truck stopped and started or turned a corner, the supply pump sucked air and caused cavitation in the high-pressure pump (and damaged the valves). The tanks also needed to be cleaned to remove sand, mud, and debris from the water supply. Cleaning sometimes involves using the Hydro-Mower separately from the truck to clean smaller areas of rubber deposits or using a handheld gun with a rotary nozzle to clean around obstacles.

Harben High-Pressure Water Technology has created simple and sophisticated systems. Generally speaking, the company's production focuses on the *Harben Century Pumps* and the *Rotoblaster*. Hydraulics or air that can be furnished by the truck systems rotates two bars. By driving the spray bars, the nozzles can be directed in the most effective

Figure 12.16. Runway cleaning in Beijing (*Courtesy of Harben*)

direction. A vacuum system is mounted at the back of the truck and a flex hose removes water and debris from the *Rotoblaster* housing. Units went to Kai Tak, Hong Kong, and to Abu Dhabi. Harben supplied systems to a contractor who works at the London airport. Harben also supplied parts/components for a system that went to Venezuela, and the contractor built his own unit. He mounted the pump units on the truck chassis and towed the water tank trailer. A unit that went to Beijing, China, had no recovery system, as shown in Fig. 12.16; another unit had the vacuum system, as shown in Fig. 12.17. These systems all work the same way, with two pump units and rotating cleaning heads.

Because the speed of the truck that contains the Harben units is so critical, it incorporates a system to disengage the road engine, and the hydraulic system drives the truck at a constant speed with a cleaning path that is 8-ft wide. The straight bar is used for apron cleaning, and this unit was used at construction sites to clean concrete. When a space is too small for a truck, a hand-operated *Rotoblaster* unit can be used. The airport at Orlando, FL, has a hand-operated *Rotoblaster* system that is used for smaller cleaning jobs.

Darrel LaSage at Vac-Con has worked with Harben in the United States to offer a vacuum system for the runway-cleaning truck (uses a Sterling truck with good hydraulics). A contractor group in the Newark,

Figure 12.17. Runway cleaning truck with vacuum system (*Courtesy of Harben*)

NJ, area wanted to assemble its own systems. The group used a White garbage truck chassis with the Harben pump units and *Rotoblaster* mounted on the truck, which also towed the water tank trailer. The group had contracts to clean the runways at Newark, LaGuardia, and other airports in the New York area.

The Harben rotary pump runs 14–20 gpm and operates well up to 8,000–10,000 psi. The runways of a military base in the Middle East (Shasha) near Dubai were covered with black asphalt, and it was difficult to see the rubber. Using water made the conditions worse, so the runway was divided into squares (marked using a yellow wax crayon), and local labor forces cleaned them by hand. The work was done during the daytime, and the tower would let the contractors know when to get off the runway. The aircrafts told the tower that they could see the clean areas (about 5 miles out) during their approach. The process was not sophisticated, but the job was completed.

As a contractor, you would need to be able to service several airports to stay in business. A complete system might run $300,000 USD, and it could be used for apron cleaning, paint line removal, and the rubber removal. The first unit that Shaun Virgo ever supplied went to Istanbul, where an airport's runway was in such poor condition that the planes were refusing to land. The airport was using a water-jetting unit with multiple guns to attempt to clean the surfaces, but that type of operation cannot be mobilized quickly. A complete unit can move off the runway very rapidly and then return to the job site to continue cleaning. Another application involves deicing planes with a methanol-based fluid, which is detrimental to the environment. Water jets can be used for immediate cleaning after use of these deicing fluids to prevent some of the harmful effects on the surrounding environment.

12.4   Automotive Plant Systems

High-pressure pump systems have been used to deburr machined parts, such as automatic transmissions, or have been used to clean sand from engine block castings in the automotive industry for a number of years. Automated systems used in manufacturing include robotic cleaning heads using 20,000-psi water jets to remove the paint buildup on vehicle paint carriers and simple nozzle arches to wash auto bodies on a production line. The paint booth grate cleaning pump systems evolved from hand-operated water-blasting guns into power rotating nozzles in lawnmower-type equipment. The high-pressure pumps, controls, and piping systems are permanently installed using sound-attenuated enclosures that have become part of the automotive plant design.

12.5   *Rubber Crumb* Conveyor Cleaner

Some rubber plants produce a product called *rubber crumb* that is later made into finished products. High-pressure water jets proved to be a good way to clean the *Rubber Crumb Dryer Belts* built by Proctor & Schwartz and others that are used in the manufacturing process. Several companies, such as Goodyear and DSM Copolymer, hired contractors who used hand lances or mechanized nozzle movement systems to do the cleaning. When it became feasible to install permanent equipment, fixed automated systems were designed and installed. One design uses a trolley track with a high-pressure swivel to move across the surfaces to be cleaned. The devices can be air powered, or just the swivel with its cleaning nozzles may have air-powered rotation. To get the continuous movement of the swivel carrier, a scotch yoke attached to a chain loop can be used to provide the back-and-forth motion for conveyor coverage.

12.6   *Cleaning Cabinets*

A product called the *cleaning cabinet* has been used to contain the debris that is being removed by a device such as a high-pressure water jet. Investment castings that are produced with tree-type pouring channels or intricate castings with a pour spue and passages are good candidates for cleaning cabinets. This type of cabinet was developed by water-blasting manufacturers in the 1970s and proved to be an economical and safe way to rapidly remove the casting material from the metal parts.

12.7   Cable Cleaning Station

The *Tow Cable Wash Station* was developed for the Westinghouse Electric Oceanic Division in 1988 to clean and rinse the cable used for "towed-array" equipment. After the cable was used in saltwater, it collected salt, marine growth, and debris, so it had to be cleaned. The wash station was designed to allow the cable to be pulled through for cleaning at a rate of 13 ft/min. The cart-mounted unit consisted of a *Hypro Series 8600* twin plunger pump, 1–½ hp electric motor, and wash chamber with self-pulsed nozzles to produce 1,000 psi at 3 gpm.

12.8   Airplane Cleaner

A prototype *Pulsed Jet Cleaner* was part of a research project for the U.S. Air Force to clean airplanes. Self-resonating nozzles were used with a *Cat 650* plunger pump and a 10-hp motor to produce 2,500 psi at 6 gpm. The portable unit was tested in 1985 on cargo plants at Robins Airforce Base near Macon, GA. For greater coverage, three nozzle lances were fitted to one shut-off style handgun. Detergent could be applied by a lower lance supplied by a separate pump at a pressure of 100 psi and flow of 2.2 gpm. The type of nozzles used in this prototype was also used to clean automotive vehicles, remove ice from the topsides of ships, and clean various surfaces. A January 1991 article in *Cleaner Times* described a hot-water washer using the pulsed jet to clean oil and dirt from the flight deck of the U.S.S. John F. Kennedy. The unit produced 3,000-psi at 5.5 gpm of water that was heated by a jet fuel-fired burner; the unit also had a wet/dry vacuum. The features allowed for good cleaning of the carrier deck and tie-down holder cups.

12.9　Erosion Material Tester

The testing of materials for their resistance to cavitation before the late 1980s was done by an American Society of Testing Materials (ASTM) vibratory erosion tester as described in the *ASTM G32-77 Annual Book of ASTM Standards*. The test involves vibrating a specimen of material in water to measure the effect of cavitation, but the test was inherently slow because of the limited power input. It was also limited by its inability to test materials such as concrete, rocks, or multilayered specimens that could not be readily mounted onto the vibrating tip of the system. Before its closing in 1988, Hydronautics developed three models of a cavitating jet erosion test apparatus (ETA). The basic concept consisted of a nozzle designed to induce the explosive growth of vapor-filled cavities within a relatively low-velocity liquid jet. The high-pressure pump size, pressure flow, and horsepower could be 5, 10, or 25 hp, where the vibratory units were typically limited to about 200 W or about 0.27 hp. The testing time was reduced, so the acquisition cost could be recovered from the reduced labor costs. Laboratories in the United States and Canada operated the ETA units in one of three sizes. Figure 12.18 shows the 25-hp unit with the high-pressure pump built into the lower cabinet and a stainless-steel

Figure 12.18. Erosion material tester unit (*G. Matusky*)

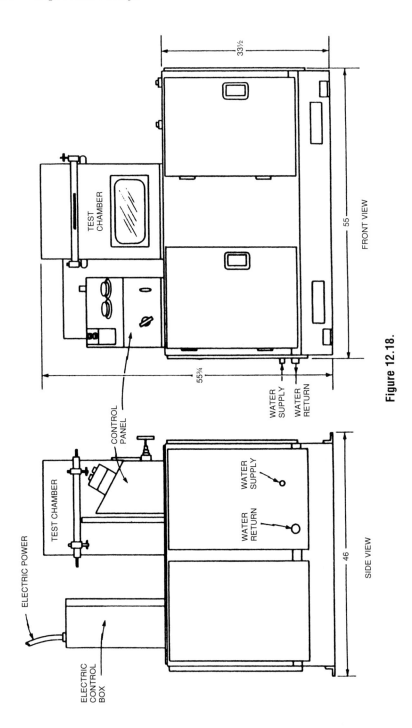

Figure 12.18.

test chamber. Tests on Stellite (nickel, cobalt, and iron alloys made by the Deloro Stellite Company), for instance, could be done in 5 min, while it took up to 30 days to accomplish the tests with a vibratory tester.

12.10 Vessel Cleaning System

The Uraca Company of Bad Urach in Germany reported that a customer in the United States, ICI Americas, was looking for an economical and environmentally safe means of cleaning an existing vessel used in the production of specialty plastics. The project required that the cleaning system meet or exceed the following minimum specifications:

- The vessel should be cleaned in less than 1 hour.
- The system should be fully automated and use the plant DCS system.
- The system should not require a manual entry into the vessel.
- The system must be leak free to avoid the release of toxic gases from the vessel.
- The system would need to demonstrate a short return on investment.
- The system had to be delivered in less than 6 months.

Previously at this company, cleaning of the vessel was done on a random basis with manual equipment. The customer used high-pressure equipment that required occasional manual entry into the vessel. This process was not only slow, but it was also hazardous to the workers cleaning the vessel. Exposure to the toxic gases and injury caused by the use of high-pressure water were constant potential dangers to those workers.

Talks were initially held with numerous potential suppliers, including Hydronautics in Maryland. The technical objective was to use existing entry openings and to maintain as much of the existing reactor pipework as possible without having to make room for the cleaning system. The customer contacted Uraca's sales and service center in the United States (Chemac) and invited the engineering team to discuss the best solution to meet the customer's demands. Detailed reviews of the internal parts of the vessel and the required cleaning program integration presented unique technical challenges. The leak-free system would be required to precisely deliver the cleaning tools inside the vessel with a knee-type delivery system via a fully automated control system.

The task was complicated somewhat by the reactor construction that had eight vertical flow interrupters (used as heating modules) inside the reactor and a four-level agitator with four blades per level. The worst-case scenario would only allow a 350-mm space between the flow interrupters and agitator blade ends. For this special application, Uraca developed a patented new cleaning system that contained the following standard components: leak-free hose reel system with approximately 24 ft of travel, a pneumatic rotating/swivel knee, a parking garage (for the cleaning head in special leak-free design), a high-pressure pump providing 8000 psi and 40 gpm of water, a Uraca tank cleaning head, and an Allen Bradley control system integrated into the plant's DCS system.

The technical and delivery demands challenged project management to coordinate the design, engineering, and delivery of this special

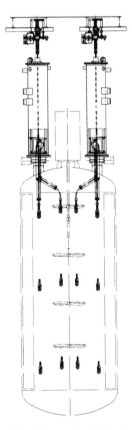

Figure 12.19. Vessel cleaning system (*Courtesy of Chemac/Uraca*)

system to the site within the required time period. The customer chose Uraca as the only supplier that met or exceeded the demands required for this project. The project required weekly coordination meetings between the customer and the engineering staff to ensure that the customer was up-to-date with each step of the design and manufacturing process.

In less than 6 months, the customer had the Uraca system in place and ready for start-up. The project was completed on time and was started up without any problems by Joseph Geronimo (of Chemac) and Otto Baechle (of Uraca). The customer was able to realize all the goals set out for the project, and the system paid for itself in the first year of operation by improving product quality and increasing production for the plant; in addition, there had been not been a need to build a new vessel because the original vessel was kept in production for a longer period of time, which also saved money for the company. Figure 12.19 is a sample drawing of the system provided to the customer.

References

ASTM G32-77 Annual Book of ASTM Standards. (1981). **Part 10**, 929–935.

Gracey, M. T. (2001). Where the rubber meets the runway. *Cleaner Times*. Little Rock, AR.

Gracey, M. T. (1997). Cleaning process equipment with automated high-pressure water. *Ninth American Water Jet Conference*, Dearborn, Michigan.

Gracey, M. T., and Conn, A. F. (1987). Research facilities using cavitating water-jet technology to test materials. *ASME Winter Meeting*, Boston, Massachusetts.

Appendix A

Corrosion Data

Corrosion Data

Chemicals	Aluminum	Brass	Carbon Steel	Ductile Iron/Cast Iron	316 Stainless Steel	17-4PH	Alloy 20	Monel	Hastelloy C	Buna N (Nitrile)	Delrin/Lubetal	EPDM/EPR	Viton	Hypolon	Neoprene	Nylon	Grafoil	Teflon - Reinforced/ or Polyfill
Acetaldehyde	B	C	C	C	A		A	A	A	D	A	B	C	D	D	B		A
Acetamine	B	B	B	B	B				A	A				B	B			A
Acetate Solvents	A	B	A	B				A	A	D	D		D	D	D			A
Acetic Acid, aerated	B	D	D	D	A			A	A	C	D		C		C	B	A	A
Acetic Acid, Air Free	B	B	D	D	A	A	A	A	A	C	D		D		C	B	A	A
Acetic Acid, crude	C	C	C	C	A	A	A	B	A	D	D		D		D	B	A	A
Acetic Acid, glacial					A			A		D		B	C	C	C	B	A	A
Acetic Acid, pure	C	C	D	D	A	A	A	A	D	A	D	D	D	A	D	B	A	A
Acetic Acid, 10%	C	C	C	C	A	A	A	B	A	D	B	B	D	C	C	B	A	A
Acetic Acid, 80%	C	C	C	C	A	A	A	B	A	D	D	C	D	D	D	B	A	A
Acetic Acid Vapors	B	D			D	D	B	C	A	D						B	A	A
Acetic Anhydride	B	D	D	D	B	B	B	B	A	D	C	C	D	B	C		A	A
Acetone	A	A	A	A	A	A	A	A	A	D	A	A	D	D	D	A	A	A
Other Ketones	A	A	A	A	A	A	A	A	A	D	A	D	D	D	D			A
Acetyl Chloride	D	A		D	C			B	A	D	D	D	D	D	D			A
Acetylene	A	B	A	A	A	A	A	A	A	B	A	A	A	C	C	A		A
Acid Fumes	B	D	D	D	B		B			C	D			C	B			A
Acrylonite	B	A	A	C	A		B	A	A	D	D	D	C	D	D	A		A
Air	A	A	A	A	A		A	A	A	A	A	A	A	A	A	A		A
Alcohol, Amyl	B	B	B	C	A		B	B	B	C	A	A	B	B	C	A	A	A
Alcohol, Butyl	B	B	B	C	A		A	A	A	B	A	C	A	B	B	A	A	A
Alcohol, Diacetone	A	A	A	A	A		A	B	A	D	A	B	D	C	C	A	A	A
Alcohol, Ethyl	B	B	B	B	B		A	B	A	A	A	A	B	B	A	A	A	A
Alcohols, Fatty	B	B	B	B	A		A		A	B	A			B	B	A	A	A
Alcohol, Isopropyl	B	B	B	B	B		A	B	B	C	A	A	A	B	B	A	A	A
Alcohol, Methyl	B	B	B	B	A		A	A	A	B	A	A	C	A	A	A	A	A
Alcohol, Propyl	A	A	B	B	A		A	A	A	B	A	A	A	B	B	A	A	A
Alumina	A	A						A	A	A	A			A	A			A
Aluminum Acetate	C	D		D	A	B	B	C	B	D	D	A	D	D	D			
Aluminum Chloride dry	B	B	C	D	C		D	B	B	B	A	A	A	B	B	A	A	A

Ratings: **A**–Excellent **B**–Good **C**–Poor **D**–Do not use **Blank**–No information

Corrosion Data

Chemicals	Aluminum	Brass	Carbon Steel	Ductile Iron/Cast Iron	316 Stainless Steel	17-4PH	Alloy 20	Monel	Hastelloy C	Buna N (Nitrile)	Delrin/Lubetal	EPDM/EPR	Viton	Hypolon	Neoprene	Nylon	Grafoil	Teflon - Reinforced/ or Polyfill
Ammonium Sulfite	C	C	C	C	A		B	D		B	A	B	A		A			A
Amyl Acetate	B	B	C	C	B	A	A	B	A	D	A	V	D	D	D	D		A
Amyl Chloride	D	B		B	A		A	B	B	D	A	D	D	D	C	B		A
Aniline	C	D	C	C	B		A	B	B	D	D	C	C	D	D	B	A	A
Aniline Dyes	C	C	C	C	A		A	A		C	A	D	B	C	C			A
Apple Juice	B	C	D	D	B		A	A		A	A	B	A	B	A			A
Aqua Regia (Strong Acid)	D	D	D	D	B		B			D	D	D	D	D	D		D	A
Aromatic Solvents	A	A	C	B	A		A	B		D	A	D		A	D			A
Arsenic Acid	D	D	D	D	B		B	D	B	A	D	B	A	B	A		A	A
Asphalt Emulsion	C	A	B	B	A		A	A	A	D	A	D	A	D	C	A		A
Asphalt Liquid	C	A	B	B	A		A	A	A	C	A	D	A	D	C			A
Barium Carbonate	C	B	B	B	B			B	B	A	B	A	A	A	A	A		A
Barium Chloride	D	B	C	C	B	B	C	B		A	A	A	A	B	A			A
Barium Cyanide	D	C		C	B		B	D		B	A	B	B	B	B			A
Barium Hydrate	D	D			A		A	B			A							A
Barium Hydroxide	D	C	C	B	B	A	A	B		A	A	B	A	B	A	A		A
Barium Nitrate	B				A		A			A				B				A
Barium Sulfate	D	C	C	C	A		A	B		A	A	B	A	B	A			A
Barium Sulfide	D	D	C	D	B		B	C		A	A	A	A	B	B	B		A
Beer	A	B	D	D	A	A	A	A		B	A	B	A	C	B	A		A
Beet Sugar Liquors	A	A	B	B	A		A	A		A	A	B	A	C	A	A		A
Benzaldehyde	A	A	A	C	A		A	B	B	D	A	A	D	D	D	A		A
Benzene (Benzol)	B	B	B	B	B	B	A	A	B	D	C	D	B	D	D	A	A	A
Benzoic Acid	B	B	D	D	B	A	B	B	A	C	A	D	B	D	C	D		A
Berryllium Sulfate	B	B		B	B		A	B		B	A	B	B	B	B			A
Bleaching Powder wet		B			C		B	D	A	D	D	B	B	B	A	D		A
Blood (Meat Juices)	B	B		D	A	A	A	B		B	A	B	B	B	B			A
Borax (Sodium Borate)	C	D	C	C	A			A	A	B	A	A	A	B	D			A
Bordeaux Mixture					A		A				A		A					A
Borax Liquors	C	A	C	C	B		A	A	B		A	A	A	D	C			A

Ratings: **A**–Excellent **B**–Good **C**–Poor **D**–Do not use **Blank**–No information

Corrosion Data

Chemicals	Aluminum	Brass	Carbon Steel	Ductile Iron/Cast Iron	316 Stainless Steel	17-4PH	Alloy 20	Monel	Hastelloy C	Buna N (Nitrile)	Delrin/Lubetal	EPDM/EPR	Viton	Hypolon	Neoprene	Nylon	Grafoil	Teflon - Reinforced/ or Polyfill
Carbon Tetrachloride, dry	B	C	B	C	A	A	A	A	A	D	A	D	B	D	D	A	A	A
Carbon Tetrachloride, wet		D	D	D	B		B	B	B	D	B	D	B	D	D	A	A	A
Casein	C	C		C	B		B	C		B	A	B	B	B	B			A
Caster Oil	A	A	B	B	A		A	A	A	A	A	B	A	B	B	A		A
Caustic Potash					A		A	N		B	D		B	B				A
Caustic Soda	D		B	B	A		A	A		C	D	B	B	B				A
Cellulose Acetate	B	B		B	B			B	B	D	C	B	D	D	D			A
China Wood Oil (Tung)	A	C	C	C	A		A	A	A	A	A	D	A	B	B	A		A
Chlorinated Solvents	D	C	C	C	A		A	B		D	A	D	C	D	D			A
Chlorinated Water	C				C	D	A	D	D	B	D		A	B	A	D	B	A
Chlorine Gas, dry	B	C	B	B	B	C	A	A	A	C	D	D	B	C	D	D	A	A
Chlorobenzene, dry	B	B	B	B	A		A	B	B	D	B	D	A	D	D	A		A
Chloroform, dry	D	B	B	C	A	B	A	A	B	D	A	D	B	D	D	B		A
Chlorophyll, dry	B	B		B	B		A	B		B		B	B	B	B			A
Chlorosulfonic Acid, dry	B	C	B	B	B		B	B	A	D	D	D	D	D	D	D		A
Chrome Alum	C	C	B	C	A		A	B		B	B	B	B	B	B	D		A
Chromic Acid<50%	C	D	D	D	C	C	B	C	B	D	D	C	C	B	D	D		A
Chromic Acid>50%	D	D	D	C	C	D	B	D	B	D	D	C	C	B	D	D		A
Chromium Sulfate	B	C		D	B		C	B		B	C	B	B	B	B			A
Cider	B				A		B	A		A			D					A
Citric Acid	B	C	D	D	B	C	A	B	A	B	A	B	A	A	A	B	A	A
Citrus Juices	C	B	D	D	B		A	A		A	A		A	D	A			A
Coca-Cola Syrup					A		A			B	A		B	D	B			A
Coconut Oil	B	B	C	C	B		A	B		A	A	A	A	D	C			A
Coffee	A	A		D	A		A	B		A	A	A	A	C	A			B
Coffee Extracts, hot	A	B	C	C	A		A	A			A					D		A
Coke Oven Gas	A	C	B	B	A		A	B		C	D	D	B	C	D	C		A
Cooking Oil	B	B	B	B	A		A	A		A	A	D	A	C	B			A
Copper Acetate	D	D	D	D	A		A	C	B	C	D	B	D	D	C			A
Copper Carbonate	D				A		A			A								A

Ratings: **A**–Excellent **B**–Good **C**–Poor **D**–Do not use **Blank**–No information

Corrosion Data

Chemicals	Aluminum	Brass	Carbon Steel	Ductile Iron/Cast Iron	316 Stainless Steel	17-4PH	Alloy 20	Monel	Hastelloy C	Buna N (Nitrile)	Delrin/Lubetal	EPDM/EPR	Viton	Hypolon	Neoprene	Nylon	Grafoil	Teflon - Reinforced/ or Polyfill
Drilling Mud	B	B	B	B	A		A	B		A	A	A	A	B	C			A
Dry Cleaning Fluids	A	C	B	B	A		A	B		D	A		B		D			A
Drying Oil	C	C	C	B	B		B	B		A	A			B				A
Enamel		A								B	A	D			B			A
Epsom Salts (MgSo)₄	A	B	C	C	B		B	B		A	A		A	D	A	B		A
Ethane	A	B	C	C	B		B	B		A	A	D	A		B			A
Ethers	A	B	A	B	A	B	A	B		D	C	C	C	D	D			A
Ethyl Acetate	A	C	B	C	B	A	B	B	B	D	C	C	D	D	D	A		A
Ethyl Acrylate	C	B	C	C	A		A	B	A	D	B	C	D	D	D	A		A
Ethyl Benzene					A		A	C	A	D			D	B				A
Ethyl Bromide	B	A		B	B		C	B		B	A	B	B	D	B			A
Ethyl Chloride, dry	B	B	B	B	A	A	A	B	B	C	A	C	B	D	C	A		B
Ethyl Chloride, wet	D	C	D	D	B		B	B	B	C	A	B	B	D	C	A		A
Ethylene Chloride	C				A		A	B	B	D	A		D		A			A
Ethylene Dichloride			B		A		B		D	C	D	D	D	D	A	A	A
Ethylene Glycol	A	B	B	B	B	A	A	B	A	A	A	A	A	B	B	A		A
Ethylene Oxide	C	C	B	B	B		B	B	A	D	A	D	D	D	D	D		A
Ethyl Ether	B	B		C	A		A	A	B	D	A	D	D	D	D			A
Ethyl Silicate		B		B	B		B	B		B	A	B	B	B	C			A
Ethyl Sulfate		B			B		B			B	A	C	A		B			A
Fatty Acids	B	C	D	D	A		A	B	A	B	A	D	A	D	B	A	A	A
Ferric Hydroxide				A			A	A		B	A							A
Ferric Nitrate	D	D	D	D	C	B	A	D	B	A	A	A	A	B	A			A
Ferric Sulfate	D	D	D	D	B	B	A	D		A	A	A	A	A	A	C		A
Ferrous Ammonium Citrate	B				B		B				A							A
Ferrous Chloride	D	B	D	D	D		D	D	D	A	A	A	A	B	A	C	A	A
Ferrous Sulfate	C	B	D	D	B		B	B	B	A	A	A	A	B	A		A	A
Ferrous Sulfate, Saturated	C	C	C	C	A		A	B	B	C	A	B	B	B	C			A
Fertilizer Solutions	B	C	B	B	B		B	B		B					B			A
Fish Oils	C	B	B	B	A		A	A		A	A	D	A	D	B			A

Ratings: A–Excellent **B**–Good **C**–Poor **D**–Do not use **Blank**–No information

Corrosion Data

Chemicals	Aluminum	Brass	Carbon Steel	Ductile Iron/Cast Iron	316 Stainless Steel	17-4PH	Alloy 20	Monel	Hastelloy C	Buna N (Nitrile)	Delrin/Lubetal	EPDM/EPR	Viton	Hypolon	Neoprene	Nylon	Grafoil	Teflon - Reinforced/ or Polyfill
Glue	A	B	A	B	B		A	B	A	A	A	B	A	B	A	A		A
Glycerine (Glycerol)	A	B	C	B	A	A	A	A	A	C	A	A	B	A	D	A	A	
Glycol Amine	C	D		B	B	A			D	A	C	D	D	C			A	
Glycol	A	B	CC	B	B		A	B		B	C	A	A	B	A			A
Graphite	B	B		C	B		A	B		B	A	B	B	B	B			A
Grease	B	C	A	A	A		A	B		A	A	D	A	D	B			A
Helium Gas	B	B		B	A		A	B	A	B	A	B	B	B	B			A
Heptane	A	A	N	N	A		A	N	A	A	A	D	A	B	B			A
Hexane	A	B	B	B	A		A	B	A	A	A	D	A	B	C	A		A
Hexanol, Tertiary	A	A	A	A	A		A	A	A	A	A	D	B	C	C	A		A
Hydraulic Oil, Petroleum Base	A	B	A	B	A		A	A		A	A	D	A		B			A
Hydrazine	C	D		D	B		B	D		C	D	B	D	C	C			A
Hydrocyanic Acid	A	D	D	C	A		A	C	B	B	D	B	A	B	B			A
Hydrofluosilicic Acid	D	A	D	D	C		B	B		B	A	B	A	B	B		A	A
Hydrogen Gas, cold	A	B	B	B	A		A	A		B	A	B	A	B	B	A		A
Hydrogen Gas, hot	C		B		B	A		A	B	A	B			B				A
Hydrogen Peroxide, Concentrated	A	D	D	D	B		B	D	D	D	D	B	B	B	B	D	D	A
Hydrogen Peroxide, Dilute	A	C	D	D	B		B	D	D	A	D	B	A		B	D		A
Hydrogen Sulfide, Dry	A	C	B	B	A	B	B	B	B	C	C	A	A	B	A	D	A	A
Hydrogen Sulfide, Wet	B	D	C	D	B		B	C	D	C	C	B	A	B	B	D	A	A
Hypo (Sodium Thiosulfate)	B	C	D	C	B		B	B		A	A	A	A		A			A
Illuminating Gas	A	A	A	A	A		A	A		C	A	D	A	D	C			A
Ink - Newsprint	C	C	D	D	A		A	B		A	A	B	A	B	B	A		A
Iodoform	C	C	B	C	A		A	C		A		A						A
Iso-Butane				B	B			B	A	D			D					A
Iso-Octane	A	A	A	B	A		A	A		A	A	D	A	B	C			A
Isopropyl Acetate				B	A				D	A	D			D			A	A
Isopropyl Ether	B	A	A	B	A		A	B	A	C	A	D	D	C	A	A	A	A
J P-4 Fuel	A	A	A	B	A		A	A	A	A		A		C	A			A
J P-5 Fuel	A	A	A	A	A		A	A	A	B	A		A	C	A			A

Ratings: **A**–Excellent **B**–Good **C**–Poor **D**–Do not use **Blank**–No information

Corrosion Data

Chemicals	Aluminum	Brass	Carbon Steel	Ductile Iron/Cast Iron	316 Stainless Steel	17-4PH	Alloy 20	Monel	Hastelloy C	Buna N (Nitrile)	Delrin/Lubetal	EPDM/EPR	Viton	Hypolon	Neoprene	Nylon	Grafoil	Teflon - Reinforced/ or Polyfill
Maleic Anhydride	B	B		B	B		B	B	B	D	C	D	B	D	D			A
Malic Acid	B	B	D	D	B		B	B		A	A		A		B			A
Malt Beverages					A		B	A		A	A	B	A		A			A
Manganese Carbonate	B				B		A		B	A								A
Manganese Sulfate	B	B		D	A		A	B		B	A	B	B	B	B		A	A
Mayonnaise	D	D	D	D	A		A	B		A	A		A		A			A
Meat Juices	B	D			A		A		B	A				B	B			A
Melamine Resins				D	C		C		B	A			D	B				A
Methanol	B	B		B	A		A	B		B	C	D	B	D	B			
Mercuric Chloride	D	D	D	D	B		B	D	B	A	A	A	A	B	B	C		A
Mercuric Cyanide	D	D	D	D	A		A	C	B	A	A	A	A	B	B			A
Mercurous Nitrate	D	D			A		A	D		A			B	B				A
Mercury	D	D	A	A	A		A	B	B	A	A	A	A	B	A			A
Methane	A	A	B	B	A		A	B	A	A	A		A		B			A
Methyl Acetate	A	A	B	B	A		A	B	A	D	B	B	D	D	D	A		A
Methyl Acetone	A	A	A	A	A		A	A		D	B	A	D	D	D	A		A
Methylamine	A	D	B	B	A		A	C	B	D	A	B	D	D	D	A		A
Methyl Bromide 100%	C	C		D	B		A	B		B	A	D	B	D	D			
Methyl Cellosolve	A	A	B	B	A		A	B	B	C	A	B	D	D	D	B		A
Methyl Cellulose					A		A		B	D	A			D				A
Methyl Chloride	D	B	B	B	A		A	B		D	A	D	B	D	D	A		A
Methyl Ethyl Ketone	A	A	A	A	A		A	A	B	D	A	B	D	D	D	A	A	A
Methylene Chloride	C	A	B	B	A		A	B	B	D	A	D	C	D	D	A		A
Methyl Formate	C	A	C	C	B		A	B	B	D	A	B	D	B	B			A
Methyl Isobutyl Ketone					A		A			D	A		D				A	A
Milk & Milk Products	A	B	D	D	A		A	B		A	A	A	A	B	A	A		A
Mineral Oils	A	B	B	B	A		A	A		A	A	D	A	C	B			A
Mineral Spirits	A	B	B	B	B		B	B		A	A		A		C			A
Mixed Acids (cold)	D	D	C	C	B		B	C		D	D	D	B	D	D	C		A
Molasses, crude	B	A	A	A	A		A	A		A	A		A	B	A	A		A

Ratings: A–Excellent B–Good C–Poor D–Do not use **Blank**–No information

Corrosion Data

Chemicals	Aluminum	Brass	Carbon Steel	Ductile Iron/Cast Iron	316 Stainless Steel	17-4PH	Alloy 20	Monel	Hastelloy C	Buna N (Nitrile)	Delrin/Lubetal	EPDM/EPR	Viton	Hypolon	Neoprene	Nylon	Grafoil	Teflon - Reinforced/or Polyfill
Oleum	B	C	B	D	B		B	C	B	D	D	D	C	B	D	D		A
Oleum Spirits	D	D		D	B		B	D		C	D	D	A	D	D	D		A
Olive Oil	B	C	B	B	A		A	A		A	A	B	A	B	B	A		A
Oxalic Acid	C	B	D	D	B	D	B	B		C	C	B	A	B	B	D	A	A
Oxygen	A	A	B	B	A	A	A	A	A	B	D	A	A	A	B	D		A
Ozone, Dry	A	A	A	A	A		A	A	D	C	A	B	A	D	D			A
Ozone, Wet	B	B	C	C	A		A	A	D	C	B	B	B	D	D			A
Paints & Solvents	A	A	A	A	A		A	A		D	A	D	B	D	D			A
Palmitic Acid	B	B	C	C	B		B	B		B	A	B	A	D	B	D		A
Palm Oil	A	B	C	C	B		A	A		B	A	D	A	D	B	A		A
Paper Pulp	D	B		B	A		A	B		B	A	B	B	B	B	A		
Paraffin	A	A	B	B	A		A	A	A	A	A	D	A	B	C	A		A
Paraformaldehyde	B	B	B	B	B		B	B		B	A	D		B				A
Paraldehyde					B		B			B	A	D		B			A	A
Pentane	A	A	B	B	A		A	B		A	A	D	A		B			A
Perchlorethylene, dry	B	C	B	B	A		A	B	B	D	B	D	A	D	D	A		A
Petrolatum (Vaseline Petroleum Jelly)	B	B	C	C	B		A	A		A	A		A		B			A
Phenol	A	B	D	D	A	B	A	A	A	D	C	D	B	D	D	D		A
Phosphate Ester 10%	D	D	A	A	A		A	A		D	A	A			A			A
Phosphoric Acid 10%	D	D	D	D	D	B	B	B	D	B	D	B	A	B	A	D	A	A
Phosphoric Acid 50% Cold	D	D	D	D	B	B	B	C	B	D	B	A	B	B	B	D	A	A
Phosphoric Acid 50% Hot	D	D	D	D	D	D	B	C	B	D	B	A	B	B	B	D	A	A
Phosphoric Acid 80% Cold	D	D	B	B	A	C	B	A			C	D	B	B	C	D	A	A
Phosphoric Acid 85% Hot	D	D	C	C	B	D	B				C	D		B	C	D	A	A
Phosphoric Anhydride	A				A		A				D	B		B	D	D	A	A
Phosphorous Trichloride	D		B	C	A		A			D	D	B	B		D	D	A	A
Phthalic Acid	B	B	C	C	B		B	A	B	C	B		A		C	A		A
Phthalic Anhydride	B	B	C	C	B		B	A	A	C	A		A		C	A		A
Picric Acid	C	C	D	D	B	C	B	D	B	C	D	B	B	B	A	B		A
Pineapple Juice	A	C	C	C	A		A	A		A	A		A	D	A			A

Ratings: A–Excellent B–Good C–Poor D–Do not use **Blank**–No information

Corrosion Data

Chemicals	Aluminum	Brass	Carbon Steel	Ductile Iron/Cast Iron	316 Stainless Steel	17-4PH	Alloy 20	Monel	Hastelloy C	Buna N (Nitrile)	Delrin/Lubetal	EPDM/EPR	Viton	Hypolon	Neoprene	Nylon	Grafoil	Teflon - Reinforced/or Polyfill
Potassium Sulfide	B	B	B	B	A		A	C	A	A	A	B	B	B	B	A		A
Potassium Sulfite	B	B	B	B	A		A	C	B	B	A	A	B	B	B	A		A
Producer Gas	B	B	B	B	B	A	B	A		A	A	D	A		B			A
Propane Gas	A	A	B	B	B	A	A	B	A	A	A	D	A	B	B	A		A
Propyl Bromide	B	B		B	B		A	B		B	A	B	B	D	B			A
Propylene Glycol	A	B	B	B	B		B	B		A	C	B	A	B	A			A
Pyridine	B			B	B		A			D	D		D		D			A
Pyrolgalic Acid	B	B	B	B	B	A	A	B		A	A		A		A			A
Quench Oil	A	B	B	B	A		A			A	A		A	B	B			A
Quinine, Sulfate, dry					A	B	A	B					A					A
Resins & Rosins	A	A	C	C	A	B	A	A		C	A		A		C	A		A
Resorcinol				B			B											A
Road Tar	A	A	A	A	A		A	A		B	A	D	A	D	C			A
Roof Pitch	A	A	A	A	A		A	A		B	A		A		C			A
Rosin Emulsion	A	B	C	C	A		A	A		D	A		B					A
R P-1 Fuel	A	A	A	A	A		A	A		B	A		A		C			A
Rubber Latex Emulsions	A	A	B	B	A		A				A		A					A
Rubber Solvents	A	A	A	A	A		A	A		D	C		D		C			A
Salad Oil	B	B	C	C	B		A	B		A	A	B	A	B	A			A
Salicylic Acid	C	C	D	D	A		B	B		A	A	B	A	B	A			A
Salt (NaCl)	B	B	C	C	B		A	A		A	A		A		A			A
Salt Brine	B	B		D	B		B	B		A	A	B	B	D	D	C		A
Sauerkraut Brine					B		B						C					A
Sea Water	C	C	D	D	B		B	A		A	A	A	A	C	A	C		A
Sewage	C	C	C	D	B	A	B	B		A	B	B	B	B	C			A
Shellac	A	A	A	B	A		A	A		A	A				A			A
Silicone Fluids	B	B		B	B		B			B	A		B	B	B			A
Silver Bromide	D				A	C	A	B			D							A
Silver Cyanide	D	D		D	A		A	B		B	D		B	B	B			A
Silver Nitrate	D	D	D	D	A		A	D		C	A	A	A	B	C			A

Ratings: A–Excellent B–Good C–Poor D–Do not use **Blank**–No information

Corrosion Data

Chemicals	Aluminum	Brass	Carbon Steel	Ductile Iron/Cast Iron	316 Stainless Steel	17-4PH	Alloy 20	Monel	Hastelloy C	Buna N (Nitrile)	Delrin/Lubetal	EPDM/EPR	Viton	Hypolon	Neoprene	Nylon	Grafoil	Teflon - Reinforced/ or Polyfill
Sodium Metasilicate Hot	B	B	D	D	A		A	A	A		A							A
Sodium Nitrate	A	B	B	B	A	B	A	B	B	C	A	B	A	B	B	A		A
Sodium Nitrate	A			B			B	C	B	C	B	A	B		D	A		A
Sodium Perborate	B	B	B	B	B	B	B	B	B	C	A	A	A	C	B			A
Sodium Peroxide	C	D	C	C	B	B	B	B	B	C	A	A	A	B	B			A
Sodium Phosphate	D	C	C	C	B	B	B	B	B	B	B	A	A	B	C	A		A
Sodium Phosphate Di-basic	D	C	C	C	B		B	B	B	A	A	A	A		A	A		A
Sodium Phosphate Tri-basic	D	C	C	C	B		B	B	B	B	A	A	A		B	A		A
Sodium Polyphosphate						B	B	B	B	B			A		B			A
Sodium Salicylte					A			A					A					A
Sodium Silicate	B	B	B	B	B		B	B		A	A	B	A	B	A	D		A
Sodium Silicate, hot	C	C	C	C	B		B	B		A	B					D		A
Sodium Sulfate	B	B	B	B	A	B	A	A		A	A	A	A	A	A			A
Sodium Sulfide	C	D	B	B	B	A	B	B		A	A	B	A	B	A	A		A
Sodium Sulfite	B	C		A	A	A	A	B	B	A	A	B	B	B	A	D		A
Sodium Tetraborate			A	A			A			A	A	B			A			A
Sodium Thiosulfate	B	C	B	C	B	A	B	B		A	A	A	A	B	A	A		A
Soybean Oil	B	B	C	C	A		A	A		A	B	B	A	D	B			A
Starch	B	B	C	C	B		A	A		A	A	C	A	B	A			A
Steam (212°)	A	A	A	A	A	A	A	B		D	D	B	C	B	D		A	A
Stearic Acid	A	C	C	C	B		B	B	A	A	A	B	A	B	C	A	A	A
Styrene	A	A	A	B	A		A	B	A	D	A	D	B	D	D	A		A
Sugar Liquids	A	A	B	B	A		A	A		A	A	B	A	D	A			A
Sugar, Syrups & Jam	B	B		C	A	A				A				D	B			A
Sulfate, Black Liquor	C	C	C	C	B	A	B	B		C	C	B	C	D	B			A
Sulfate, Green Liquor	B	C	C	C	B	A	B	B		C	A		C	D	B			A
Sulfate, White Liquor	B	C	C	C	B	B	D	C		C	D		C	D	B			A
Sulfur	A	D	C	C	B		A	B		D	A	B	B	B	C	C		A
Sulfur Chlorides	D	B	D	D	D		A	B		D	A	C	A	B	D		A	A
Sulfur Dioxide, dry	A	B	B	B	A	A	B	B	A	D	A	A	A	D	D	A	A	A

Ratings: **A**–Excellent **B**–Good **C**–Poor **D**–Do not use **Blank**–No information

Corrosion Data

Chemicals	Aluminum	Brass	Carbon Steel	Ductile Iron/Cast Iron	316 Stainless Steel	17-4PH	Alloy 20	Monel	Hastelloy C	Buna N (Nitrile)	Delrin/Lubetal	EPDM/EPR	Viton	Hypolon	Neoprene	Nylon	Grafoil	Teflon - Reinforced/ or Polyfill
Vinyl Acetate	B	B		B	B		B	B	A		D	A		B	B	A	A	A
Water, Distilled	A	A	D	D	A	A	A	A	A	C	A	B	A	B	B			A
Water, Fresh	A	A	C	C	A	A	A	A	A	C	A	B	A	A	B	C		A
Water, Acid Mine	D	D	D	D	B	B		D	C	B	A	A	D	C	A		A	
Waxes	A	A	A	A	A		A	A	A	A	A	C	A	B	B			A
Whiskey & Wines	D	B	D	D	A		A	A	A	B	A	A	A	C	B	A		A
Xylene (Xylol), Dry	A	A	B	B	A		A	A	A	D	A	D	B	D	D	A	A	A
Zinc Bromide	D	B		D	B		B	B	A	B	A	B	B	B	B			A
Zinc Hydrosulfite	D	C	A	B	A		A	B	A	A	A	A	A		A			A
Zinc Sulfate	D	B	D	D	B		A	B	A	A	A	A	A	B	A		A	A

Ratings: A–Excellent **B**–Good **C**–Poor **D**–Do not use **Blank**–No information

Appendix B

Viscosity & Specific Gravity of Common Liquids

Viscosity & Specific Gravity of Common Liquids

Liquid	Specific Gravity	Viscosity S.S.U.		
		70° F.	100° F.	130° F.
Sugar, Syrups, Molasses, etc.				
Corn Syrups	1.4–1.47	–	5,000–500,000	1,500–60,000
Glucose	1.35–1.44	–	35,000–100,000	10,000–13,000
Honey (Raw)	–	–	340	–
Molasses	1.40–1.49	–	1,300–250,000	700–75,000
Sugar Syrups 60 Brix	1.29	230	92	–
Sugar Syrups 62 Brix	1.30	310	111	–
Sugar Syrups 64 Brix	1.31	440	148	–
Sugar Syrups 66 Brix	1.33	650	195	–
Sugar Syrups 68 Brix	1.34	1000	275	–
Sugar Syrups 70 Brix	1.35	1650	400	–
Sugar Syrups 72 Brix	1.36	2700	640	–
Sugar Syrups 74 Brix	1.38	5500	1100	–
Sugar Syrups 76 Brix	1.39	10000	2000	–
Corn Starch 22 Baume	1.18	150	130	–
Corn Starch 24 Baume	1.20	600	440	–
Corn Starch 25 Baume	1.21	1400	800	–
Ink–Printers	1.0–1.38	–	2,500–10,000	1,100–3,000
Ink–Newspaper	–	–	5,500–8,000	2400
Tallow	.918	56 SSU at 212°F.		

Continued

Viscosity & Specific Gravity of Common Liquids—Continued

Liquid	Specific Gravity	Viscosity S.S.U.		
		70° F.	100° F.	130° F.
Tars				
Coke Oven–Tar	1.12+	3,000–8,000	650–1,400	—
Gas House–Tar	1.16–1.3	15,000–300,000	2,000–20,000	—
Crude Oils				
Texas, Oklahoma	.81–916	100–700	34–210	—
Wyoming, Montana	.86–88	100–1100	46–320	—
California	.78–92	100–4500	34–700	—
Pennsylvania	.8–.85	100–200	38–86	—
Glycol				
Propylene	1.038	240.6	—	—
Triethylene	1.125	185.7	—	—
Diethylene	1.12	149.7	—	—
Ethylene	1.125	88.4	—	—
Glycerine (100%)	1.26	2900	813	—
Phenol (Carbolic Acid)	.95–1.00	60	—	—
Silicate of Soda	—	—	356–640	—
Sulfuric Acid (100%)	1.83	75	—	—

Viscosity Conversion Table

SSU Seconds Saybolt Universal	SSF Seconds Saybolt Furol	Absolute Viscosity Centipoises	Seconds Redwood (Standard)
31		1.00	29
35		2.56	32.1
40		4.30	36.2
50		7.40	44.3
60		10.20	52.3
70	12.95	12.83	60.9
80	13.70	15.35	69.2
90	14.44	17.80	77.6
100	15.24	20.20	85.6
150	19.30	31.80	128
200	23.5	43.10	170
250	28.0	54.30	212
300	32.5	65.40	254
400	41.9	87.60	338
500	51.6	110.0	423
600	61.4	132	508
700	71.1	154	592
800	81.0	176	677
900	91.0	198	762
1000	100.7	220	896
1500	150	330	1270
2000	200	440	1690
2500	250	550	2120
3000	300	660	2540
4000	400	880	3380
5000	500	1100	4230
6000	600	1320	5080
7000	700	1540	5920
8000	800	1760	6770
9000	900	1980	7620
10000	1000	2200	8460
15000	1500	3300	13700
20000	2000	4400	18400

Continued

Viscosity Conversion Table—Continued

$$\text{Kinematic Viscosity Centistokes} = \frac{\text{Absolute Visc. (Centipoise)}}{\text{Specific Gravity}}$$

$$1 \text{ Centistoke} = \frac{\text{Stoke}}{100}$$

$$1 \text{ Centipoise} = \frac{\text{Poise}}{100}$$

1 Stoke = 100 Centistokes

1 Poise = 100 Centipoises

Centipoises

The term "Centipoises" is referred to commonly as a measure of Absolute Viscosity. Convert centipoises to centistokes by dividing by the Specific Gravity of the solution at the operating temperature.

NOTE: Plotting Viscosity = If viscosity is known at any two temperatures, the viscosity at other temperatures can be obtained by plotting the viscosity against temperature in degrees fahrenheit on log paper. The point lie in a straight line.

Heat Transfer Liquids

Liquid	Pump Constr.	Seal & Gasket Matl.	Viscosity (SSU)	Specific Gravity	Vapor Pressure (PSIA)	Boiling Point (@14.7 PSIA)	Remarks
Ucon 275 CP	BF Al	Buna N Viton Teflon	275 @ 38°C 100°F	1.08 @ 20°C 68°F			
Ucon 300 CP	BF Al	Buna N Viton Teflon	200 @ 38°C 100°F				
Ucon 150 LT	BF Al	Buna N Viton Teflon	150 @ 38°C 100°F	1.08 @ 16°C 60°F			
Ucon 200 LT	BF Al	Buna N Viton Teflon	200 @ 38°C 100°F	1.08 @ 16°C 60°F			
Ucon 275 LT	BF Al	Buna N Viton Teflon	275 @ 38°C 100°F	1.08 @ 16°C 60°F			
Ucon 300 LT	BF Al	Buna N Viton Teflon	300 @ 38°C 100°F	1.08 @ 25°C 77°F			
Ucon 500	BF Al	Buna N Viton Teflon	280 @ 38°C 100°F	1.02 @ 38°C 100°F			

Estimating the Output from a Pipe

By the Horizontal Open Discharge Method

Construct an L-shaped gauge like that shown above, with the short leg 4 inches long. Make the long leg to suit the pipe sizes and capacities for which the gauge will be used (refer to table), and mark it in inches.

Lay the gauge along the top of the pipe with the short leg barely touching the stream of water, and note distance A. Read the discharge rate from the table.

EXAMPLE

$D = 3''$; $A = 15''$; $Q = 183$ gpm

Table is based on formula:

$$Q = 1.28 \times A \times (D)^2$$

Nominal Size of Pipe (D)

Discharge Rate (Q) – Gallons per Minute

A Inches	1″	1.25″	1.5″	2″	2.5″	3″	4″	5″	6″	8″	10″	12″	Average Velocity ft/sec.
4	5.7	9.8	13.3	22.0	31.3	48.5	83.5						2.1
5	7.1	12.2	16.6	27.5	39.0	61.0	104	163					2.6
6	8.5	14.7	20.0	33.0	47.0	73.0	125	195	285				3.1
7	10.0	17.1	23.2	38.5	55.0	85.0	146	228	334	580			3.7
8	11.3	19.6	26.5	44.0	62.5	97.5	166	260	380	655	1060		4.2
9	12.8	22.0	29.8	49.5	70.0	110	187	293	430	750	1190	1660	4.7

10	14.2	24.5	33.2	55.5	78.2	122	208	326	476	830	1330	1850	5.3
11	15.6	27.0	36.5	60.5	86.0	134	229	360	525	915	1460	2020	5.8
12	17.0	29.0	40.0	66.0	94.0	146	250	390	570	1000	1600	2200	6.2
13	18.5	31.5	43.0	71.5	102	158	270	425	620	1080	1730	2400	6.9
14	20.0	34.0	46.5	77.0	109	170	292	456	670	1160	1860	2590	7.4
15	21.3	36.3	50.0	82.5	117	183	312	490	710	1250	2000	2780	7.9
16	22.7	39.0	53.0	88.0	125	196	334	520	760	1330	2120	2960	8.4
17		41.5	56.5	93.0	133	207	355	550	810	1410	2260	3140	9.1
18			60.0	99.0	144	220	375	590	860	1500	2390	3330	9.7
19				110	148	232	395	620	910	1580	2520	3500	10.4
20					156	244	415	650	950	1660	2660	3700	10.6
21						256	435	685	1000	1750	2800	3890	11.4
22							460	720	1050	1830	2920	4060	11.8
23								750	1100	1910	3060	4250	12.4
24									1140	2000	3200	4440	13.0

Properties of Water

Temp. °F.	Absolute Vapor Pressure Psi.	Absolute Vapor Pressure Ft. Water	Specific Gravity (Water at 39.2°F = 1.000)	Temp. °F.	Absolute Vapor Pressure Psi.	Absolute Vapor Pressure Ft. Water	Specific Gravity (Water at 39.2°F = 1.000)
60	0.26	0.59	0.999	175	6.71	15.9	0.972
70	0.36	0.89	0.998	176	6.87	16.3	0.972
80	0.51	1.2	0.997	177	7.02	16.7	0.971
85	0.60	1.4	0.996	178	7.18	17.1	0.971
90	0.70	1.6	0.995	179	7.34	17.4	0.971
100	0.95	2.2	0.993	180	7.51	17.8	0.970
110	1.27	3.0	0.991	181	7.68	18.3	0.970
120	1.69	3.9	0.989	182	7.85	18.7	0.970
130	2.22	5.0	0.986	183	8.02	19.1	0.969
140	2.89	6.8	0.983	184	8.20	19.5	0.969
150	3.72	8.8	0.981	185	8.38	20.0	0.969
151	3.81	9.0	0.981	186	8.57	20.4	0.968
152	3.90	9.2	0.980	187	8.76	20.9	0.968
153	4.00	9.4	0.980	188	8.95	21.4	0.968
154	4.10	9.7	0.979	189	9.14	21.8	0.967
155	4.20	9.9	0.979	190	9.34	22.3	0.967
156	4.31	10.1	0.979	191	9.54	22.8	0.966
157	4.41	10.4	0.978	192	9.75	23.3	0.966
158	4.52	10.7	0.978	193	9.96	23.8	0.965
159	4.63	10.9	0.978	194	10.17	24.3	0.965
160	4.74	11.2	0.977	195	10.38	24.9	0.965
161	4.85	11.5	0.977	196	10.60	25.4	0.964
162	4.97	11.7	0.977	197	10.83	25.9	0.964
163	5.09	12.0	0.976	198	11.06	26.6	0.963
164	5.21	12.3	0.976	199	11.29	27.1	0.963
165	5.33	12.6	0.976	200	11.53	27.6	0.963
166	5.46	12.9	0.975	201	11.77	28.2	0.962
167	5.59	13.3	0.975	202	12.01	28.8	0.962
168	5.72	13.6	0.974	203	12.26	29.4	0.962
169	5.85	13.9	0.974	204	12.51	30.0	0.961
170	5.99	14.2	0.974	205	12.77	30.6	0.961
171	6.13	14.5	0.973	206	13.03	31.2	0.960
172	6.27	14.9	0.973	207	13.30	32.0	0.960
173	6.42	15.2	0.973	208	13.57	32.6	0.960
174	6.56	15.6	0.972	209	13.84	33.2	0.959

Continued

Properties of Water—Continued

Temp. °F.	Absolute Vapor Pressure		Specific Gravity (Water at 39.2°F = 1.000)	Temp. °F.	Absolute Vapor Pressure		Specific Gravity (Water at 39.2°F = 1.000)
	Psi.	Ft. Water			Psi.	Ft. Water	
210	14.12	33.9	0.959	239	24.53	59.8	0.948
211	14.41	34.6	0.958	240	24.97	61.0	0.947
212	14.70	35.4	0.958	241	25.43	62.1	0.947
213	14.99	36.2	0.957	242	25.89	63.3	0.946
214	15.29	37.0	0.957	243	26.36	64.5	0.946
215	15.59	37.0	0.957	244	26.83	65.6	0.946
216	15.90	37.7	0.956	245	27.31	66.8	0.945
217	16.22	38.4	0.956	250	29.83	73.2	0.943
218	16.54	39.2	0.956	260	35.44	87.4	0.938
219	16.86	40.0	0.955	270	41.87	103.6	0.933
220	17.19	40.8	0.955	280	49.22	122.8	0.927
221	17.52	41.6	0.955	290	57.57	144.0	0.923
222	17.86	42.5	0.954	300	67.0	168.6	0.918
223	18.21	43.3	0.954	310	77.7	197.0	0.913
224	18.56	44.2	0.953	320	89.7	228.4	0.908
225	18.92	45.0	0.953	330	103.0	264.0	0.902
226	19.28	45.9	0.953	340	118.0	305.0	0.896
227	19.65	46.8	0.952	350	134.6	349.0	0.891
228	20.02	47.7	0.952	360	153.0	399.2	0.886
229	20.04	48.6	0.951	380	195.8	517.7	0.874
230	20.78	49.5	0.951	400	247.3	663.9	0.861
231	21.17	50.5	0.951	420	308.8	842.4	0.847
232	21.57	51.4	0.950	440	381.6	1058.5	0.833
233	21.97	52.5	0.950	460	466.9	1318.0	0.818
234	22.38	53.5	0.950	480	566.1	1630.5	0.802
235	22.80	54.5	0.949	500	680.8	2000.1	0.786
236	23.22	56.6	0.949	520	812.4	2445.5	0.767
237	23.65	57.8	0.948	540	962.5	2980.4	0.746
238	24.09	58.8	0.948				

Atmospheric Pressure and Boiling Point of Water
at Various Altitudes

Altitude (feet)	Barometer Inches Mercury	Atmospheric Pressure PSIA	(ft. water)	Boiling Point °F.
−1000	31.0	15.2	32.5	213.8
−500	30.5	15.0	34.6	212.9
0.0	29.9	14.7	33.9	212.0
+500	29.4	14.4	33.3	211.1
+1000	28.9	14.2	32.8	210.2
+1500	28.3	13.9	32.1	209.3
+2000	27.8	13.7	31.5	208.4
+2500	27.3	13.4	31.0	207.4
+3000	26.8	13.2	30.4	206.5
+3500	26.3	12.9	29.8	205.6
+4000	25.8	12.7	29.2	204.7
+4500	25.4	12.4	28.8	203.8
+5000	24.9	12.2	28.8	202.9
+5500	24.4	12.0	27.6	201.9
+6000	24.0	11.8	27.2	201.0
+6500	23.5	11.5	26.7	200.1
+7000	23.1	11.3	26.2	199.2
+7500	22.7	11.1	25.7	198.3
+7000	22.2	10.9	25.3	197.4
+8500	21.8	10.7	24.7	196.5
+9000	21.4	10.5	24.3	195.5
+9500	21.0	10.3	23.8	194.6
+10000	20.6	10.1	23.4	193.7
+15000	16.9	8.3	19.2	184.0

Vacuum Conversion Data

Vacuum Inches, Mercury	Equivalents			
	PSIG	Inches of Water	Feet of Water	PSIA
30	14.7	407.70	33.90	0.00
29	14.21	393.60	32.80	.49
28	13.72	380.05	31.60	.98
27	13.23	366.40	30.50	1.47
26	12.74	352.90	29.40	1.96
25	12.25	339.30	28.30	2.45
24	11.76	325.70	27.10	2.94
23	11.27	312.10	26.00	3.43
22	10.78	298.60	24.90	3.92
21	10.29	285.06	23.70	4.41
20	9.80	271.40	22.60	4.90
19	9.31	257.80	21.50	5.39
18	8.82	244.30	20.30	5.88
17	8.33	230.70	19.20	6.37
16	7.84	217.70	18.00	6.86
15	7.35	203.60	17.00	7.35
14	6.86	190.02	15.80	7.84
13	6.37	176.40	14.70	8.33
12	5.88	162.80	13.50	8.82
11	5.39	149.30	12.40	9.31
10	4.90	135.70	11.30	9.80
9	4.41	122.10	10.20	10.29
8	3.92	108.50	9.00	10.78
7	3.43	95.00	7.90	11.27
6	2.94	81.40	6.80	11.76
5	2.45	67.80	5.70	12.25
4	1.96	54.20	4.50	12.74
3	1.47	40.70	3.40	18.23
2	.98	27.10	2.30	13.72
1	.49	13.60	1.13	14.21
0	0.00	0.00	0.00	14.7

Conversion Table – Celsius/Fahrenheit

Degrees C	Degrees F	Degrees C	Degrees F	Degrees C	Degrees F	Degrees C	Degrees F	Degrees C	Degrees F	Degrees C	Degrees F	Degrees C	Degrees F	Degrees C	Degrees F
0	32.0	13	56.4	26	78.8	39	102.2	52	125.6	65	149.0	78	172.4	91	195.8
1	33.8	14	57.2	27	80.6	40	104.0	53	127.4	66	150.8	79	174.2	92	197.6
2	35.6	15	59.0	28	82.4	41	105.8	54	129.2	67	152.6	80	176.0	93	199.4
3	37.4	16	60.8	29	84.2	42	107.6	55	131.0	68	154.4	81	177.8	94	201.2
4	39.2	17	62.6	30	86.0	43	109.4	56	132.8	69	156.2	82	179.6	95	203.0
5	41.0	18	64.4	31	87.8	44	111.2	57	134.6	70	158.0	83	181.4	96	204.8
6	42.8	19	66.2	32	89.6	45	113.0	58	136.4	71	159.8	84	183.2	97	206.6
7	44.6	20	68.0	33	91.4	46	114.8	59	138.2	72	161.6	85	185.0	98	208.4
8	46.4	21	69.8	34	93.2	47	116.6	60	140.0	73	163.4	86	186.8	99	210.2
9	48.2	22	71.6	35	95.0	48	118.4	61	141.8	74	165.2	87	188.6	100	212.0
10	50.0	23	73.4	36	96.8	49	120.2	62	143.6	75	167.0	88	190.4	–	–
11	51.8	24	75.2	37	98.6	50	122.0	63	145.4	76	168.8	89	192.2	–	–
12	53.6	25	77.0	38	100.4	51	123.8	64	147.2	77	170.6	90	194.0	–	–

Fraction/Decimal/Metric Equivalents Chart

Fraction	Decimal	Metric	Fraction	Decimal	Metric
1/64	.0156	.39688	33/64	.5156	13.09688
1/32	.0312	.79375	17/32	.5312	13.49375
3/64	.0468	01.19063	35/64	.5468	13.89063
1/16	.0625	01.58750	9/16	.5625	14.28750
5/64	.0781	01.98438	37/64	.5781	14.68438
3/32	.0937	02.38125	19/32	.5937	15.08125
7/64	.1093	02.77813	39/64	.6093	15.47813
1/8	.125	03.17500	5/8	.625	15.87500
9/64	.1406	03.57188	41/64	.6406	16.27188
5/32	.1562	03.96875	21/32	.6562	16.66875
11/64	.1718	04.36563	43/64	.6718	17.06563
3/16	.1875	04.76250	11/16	.6875	17.46250
13/64	.2031	05.15938	45/64	.7031	17.85938
7/32	.2187	05.55625	23/32	.7187	18.25625
15/64	.2343	05.95313	47/64	.7343	18.65313
1/4	.25	6.35000	3/4	.75	19.05000
17/64	.2656	6.74688	49/64	.7656	19.44688
9/32	.2812	7.17375	25/32	.7812	19.84375
19/64	.2968	7.54063	51/64	.7968	20.24063
5/16	.3125	7.93750	13/16	.8125	20.63750
21/64	.3281	8.33438	53/64	.8281	21.03438
11/32	.3437	8.73125	27/32	.8437	21.43125
23/64	.3593	9.12813	55/64	.8593	21.83813
3/8	.375	9.52500	7/8	.875	22.22500
25/64	.3906	9.92188	57/64	.8906	22.62187
13/32	.4062	10.31875	29/32	.9062	23.01875
27/64	.4218	10.71563	59/64	.9218	23.41563
7/16	.4375	11.11250	15/16	.9375	23.81250
29/64	.4531	11.50938	61/64	.9531	24.20937
15/32	.4687	11.90625	31/32	.9687	24.60625
31/64	.4843	12.30313	63/64	.9843	25.00312
1/2	.5	12.70000	1	1.0000	25.40000

Metric equivalents are in centimeters

FMC
Allowable Nozzle Loads for Reciprocating Pumps

The following chart shows the maximum allowable nozzle loads for a reciprocating pump on or about point "A" (see diagram below). Nozzle loads are composed of a resultant force (F) and resultant moment (M). Exceeding the limits below may cause mechanical failure and/or reduce the life of the pump.

An ideal piping system would be designed and supported so that the nozzle loads at point "A" would be reduced to zero.

Nominal Pipe Size of Discharge Opening	$F = \sqrt{Fx^2 + Fy^2 + Fz^2}$ (Force in Lbs.)	$M = \sqrt{Mx^2 + My^2 + Mz^2}$ (Moments in Ft. - Lbs.)
1/2	342	16
3/4	454	32
1	673	53
1–1/4	908	79
1–1/2	1089	105
2	1460	210
2–1/2	1953	315
3	2342	473
3–1/2	2730	604
4	3119	814

NOTE: The above figures are based on an operating temperature range of −20°F to 120°F. For conditions different than these please consult factory.

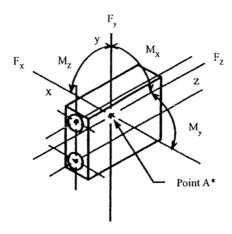

Point "A" is located at the center of the middle cylinder.

Utex Material Recommendation Chart

UTEX MATERIAL RECOMMENDATION CHART

CHEMICAL MEDIA DESCRIPTION	POLYMER TYPE	Reciprocating / Plunger Pump MOLDED (LUBRICATED) JFD	SF	SSF	ASF	BRAIDED ADJUSTABLE	BRAIDED NON-ADJUSTABLE LUBE	NO LUBE	Centrifugal / Rotary BRAIDED PREMIUM	STANDARD	MOLDED	Valve BRAIDED PREMIUM	STANDARD	Plastic Components	Metal Components
acetaldehyde	PTFE	NR	4511	4516	1650	0250	0217	0212	0214	0279	6677,4510	0686-0226	0231	PTFE	BRONZE
acetic acid, glacial	EPDM	NR	6665	1081	1610	0255	0217	0242	0226	0231	6665	0686-0226	0232	PEEK	316 SS
hot	PTFE	NR	4511	4516	1650	0250	0217	0212	0226	0210	6677,4510	0686-0226	0232	PTFE	316 SS
5%	PTFE	NR	4511	4516	1650	0250	0217	0212	0226	0210	6677,4510	0686-0226	0232	PTFE	316 SS
acetic anhydride	PTFE	NR	4511	4516	1650	0250	0217	0217	0214	0279	6677,4510	0686-0226	0226	PTFE	316 SS
acetone	EPDM	NR	6665	1081	1610	0268	0217	0242	0226	0231	6665	0686-0226	0232	PEEK	BRONZE
acetone cyanohydrin	EPDM	NR	6665	1081	1610	0268	0217	0242	0226	0231	6665	0686-0226	0232	PEEK	BRONZE
acetyl chloride	FKM	NR	6625	1073	1608	0268	0232	0242	0226	0231	6625	0686-0226	0232	PEEK	316 SS
acetylene gas	NBR	0838	0805	1056	1601	0268	0217	0242	0279	0208	0805	0686-0226	0217	PEEK	316 SS
acrylonitrile	PTFE	NR	4511	4516	1650	0250	0217	0212	0214	0279	6677,4510	0686-0226	0231	PTFE	BRONZE
adipic acid	NBR	0858	0758	1058	1658	0268	0217	0242	0226	0231	0809	0686-0226	0220	PEEK	BRONZE
alcohol, denatured	NBR	0838	0805	1056	1601	0268	0217	0242	0210	0231	0805	0686-0226	0217	PEEK	BRONZE
alkyl benzene	FKM	NR	6625	1073	1608	0268	0232	0242	0226	0231	6625	0686-0226	0232		
alkyl-arylsulphonic acid	EPDM	NR	6665	1081	1610	0268	0217	0242	0226	0228	6665	0686-0226	0226		
alkylate	NBR	0858	0758	1058	1658	0268	0217	0242	0229	0279	0805	0686-0226	0220	PEEK	BRONZE
alumina trihydrate	NBR	0858	0758	1058	1658	0268	0217	0242	0229	0279	0805	0686-0226	0217	PEEK	316 SS
aluminum acetate	FEPM	NR	6618	1043	1606	0268	0217	0242	0210	0238	6618	0686-0226	0226		
aluminum chloride	NBR	0858	0758	1058	1658	0255	0217	0242	0226	0210	0805	0686-0226	0226	PEEK	NI-RESIST
aluminum nitrate	NBR	0858	0758	1058	1658	0268	0217	0242	0279	0208	0805	0686-0226	0220	PEEK	BRONZE
aluminum potassium sulfate	NBR	0858	0758	1058	1658	0268	0217	0242	0228	0279	0805	0686-0226	0220	PEEK	BRONZE
aluminum sulfate	NBR	0858	0758	1058	1658	0268	0217	0242	0229	0279	0805	0686-0226	0217	PEEK	BRONZE
amines, mixed	FEPM	NR	6618	1043	1606	0268	0217	0242	0226	0210	6618	0686-0226	0217	PPS	CAST IRON
ammonia, gas, cold	CR	NR	6686	1066	1633	0268	0217	0242	0210	0238	6686	0686-0226	0226	PEEK	316 SS
gas, hot	FEPM	NR	6618	1043	1606	0268	0217	0242	0226	0210	6618	0686-0226	0226	PPS	NI-RESIST
liquid (anhydrous)	FEPM	NR	6618	1043	1606	0268	0217	0242	0210	0238	6618	0686-0226	0226	PEEK	316 SS
ammonium acetate	FKM	NR	6625	1073	1608	0268	0232	0242	0226	0231	6625	0686-0226	0232	PEEK	316 SS
ammonium bicarbonate	FEPM	NR	6618	1043	1606	0268	0217	0242	0210	0238	6618	0686-0226	0226	PEEK	316 SS
ammonium bifluoride	NBR	0858	0758	1058	1658	0255	0217	0242	0210	0238	0805	0686-0226	0226	PEEK	316 SS
ammonium bromide	EPDM	NR	6665	1081	1610	0268	0217	0242	0210	0238	6665	0686-0226	0226	PEEK	BRONZE
ammonium carbonate	CR	NR	6686	1066	1633	0268	0217	0242	0210	0238	6686	0686-0226	0226	PTFE	316 SS
ammonium chloride	NBR	0858	0758	1058	1658	0268	0217	0242	0210	0238	0805	0686-0226	0217	PPS	BRONZE
ammonium hydroxide	FEPM	NR	6618	1043	1606	0268	0217	0242	0229	0279	6686	0686-0226	0217	PEEK	316 SS
ammonium nitrate	NBR	0858	0758	1058	1658	0255	0217	0242	0229	0279	0805	0686-0226	0217	PEEK	316 SS
ammonium phosphate	NBR	0858	0758	1058	1658	0268	0217	0242	0229	0279	0805	0686-0226	0217	PEEK	316 SS
ammonium stearate	NBR	0858	0758	1058	1658	0268	0217	0242	0229	0279	0805	0686-0226	0217	PEEK	BRONZE

NR: Not Recommended
Consult engineering on blank spaces.
NOTE: Under JFD Design 0838 - Temperatures over 180oF; Pressure over 2500 psi UTEX recommends 0858

CHEMICAL MEDIA DESCRIPTION	POLYMER TYPE	Reciprocating / Plunger Pump MOLDED (LUBRICATED) JFD	SF	SSF	ASF	BRAIDED ADJUSTABLE	BRAIDED NON-ADJUSTABLE LUBE	NO LUBE	Centrifugal / Rotary BRAIDED PREMIUM	STANDARD	MOLDED	Valve BRAIDED PREMIUM	STANDARD	MOLDED	Plastic Components	Metal Components
ammonium sulfate	NBR	0858	0758	1058	1658	0268	0217	0242	0229	0279	0805	0686-0226	0217	0805	PEEK	BRONZE
ammonium thiocyanate	CR	NR	6686	1066	1633	0268	0217	0242	0229	0279	6686	0686-0226	0217	6686	PEEK	316 SS
amyl acetate	EPDM	NR	6665	1081	1610	0268	0217	0242	0229	0279	6665	0686-0226	0217	6665	PEEK	BRONZE
amyl alcohol	FEPM	NR	6618	1043	1606	0268	0212	0242	0279	0231	6618	0686-0226	0220	6618	PEEK	BRONZE
amyl nitrate	FEPM	NR	6618	1043	1606	0268	0212	0242	0214	0279	6618	0686-0226	0220	6618	PEEK	BRONZE
aniline	FEPM	NR	6618	1043	1606	0268	0212	0242	0214	0279	6618	0686-0226	0220	6618	PEEK	BRONZE
aniline hydrochloride	FEPM	NR	6618	1043	1606	0255	0212	0242	0214	0279	6618	0686-0226	0220	6618	PPS	NI-RESIST
anti-freeze, alcohol or glycol based	FEPM	NR	6618	1043	1606	0268	0212	0242	0214	0279	6618	0686-0226	0220	6618	PEEK	BRONZE
aqua regia	FKM	NR	6625	1073	1608	0255	0217	0242	0226	0231	6625	0686-0226	0238	6625	PEEK	NI-RESIST
argon gas	EPDM	NR	6665	1081	1610	0268	0217	0242	0279	0208	6665	0686-0226	0217	6665	PEEK	BRONZE
aroclor	FEPM	NR	6618	1043	1606	0268	0217	0242	0214	0279	6618	0686-0226	0220	6618	PTFE	NI-RESIST
arsenic acid	NBR	0858	0758	1058	1658	0268	0217	0242	0226	0231	0809	0686-0226	0238	0809	PEEK	BRONZE
ash slurry	NBR	0842	0809	1068	1604	0255	0249	0242	0226	0214	6618	0686-0226	0226	6618	PEEK	BRONZE
asphalt	FEPM	NR	6618	1043	1658	0268	0217	0242	0214	0279	6618	0686-0226	0220	6618	PEEK	BRONZE
barium chloride	NBR	0858	0758	1058	1658	0255	0217	0242	0228	0279	0805	0686-0226	0220	0805	PEEK	BRONZE
barium hydroxide; mono-, octa-, pentahydrate	NBR	0858	0758	1058	1658	0255	0217	0242	0279	0208	0805	0686-0226	0220	0805	PEEK	316 SS
barium nitrate	NBR	0858	6681	1077	1609	0268	0217	0242	0228	0279	6681	0686-0226	0220	6681	PEEK	316 SS
beer	NBR	NR	6681	1077	1609	0250	0217	0242	0226	0231	6681	0686-0226	0245	6681	PEEK	BRONZE
beet sugar	EPDM	NR	6665	1073	1608	0250	0217	0242	0226	0208	6665	0686-0226	0245	6665	PPS	BRONZE
benzaldehyde	FKM	NR	6665	1081	1608	0255	0217	0242	0279	0231	6665	0686-0226	0217	6665	PEEK	316 SS
benzene	FEPM	NR	6625	1073	1608	0250	0232	0242	0226	0231	6625	0686-0226	0232	6625	PEEK	BRONZE
benzenesulfonic acid	FEPM	NR	6618	1043	1606	0268	0217	0242	0226	0279	6618	0686-0226	0232	6618	PEEK	BRONZE
benzochloride	EPDM	NR	6618	1081	1610	0268	0217	0242	0226	0279	6665	0686-0226	0232	6665	PEEK	BRONZE
benzoic acid	FEPM	NR	6618	1043	1606	0250	0217	0242	0226	0279	6618	0686-0226	0232	6618	PEEK	BRONZE
benzotrifluoride	PTFE	NR	4511	4516	1650	0250	0217	0212	0228	0279	6677, 4510	0686-0226	0232	6677,4510	PTFE	BRONZE
black liquors	FEPM	NR	6625	1073	1606	0268	0217	0242	0226	0279	6618	0686-0226	0248	6618	PEEK	316 SS
bleach solutions	EPDM	NR	6665	1081	1610	0268	0217	0242	0229	0279	6665	0686-0226	0220	6665	PEEK	BRONZE
blood	EPDM	NR	6665	1081	1606	0250	0217	0242	0245	0245	6665	0686-0226	0245	6665	PEEK	316 SS
Bonderite® (Occidental Chemical)	NBR	0858	0758	1058	1658	0268	0217	0242	0228	0279	0805	0686-0226	0220	0805	PEEK	BRONZE
bone oil	NBR	0838	0805	1056	1601	0268	0249	0242	0279	0231	0809	0686-0226	0238	0809	PPS	316 SS
boric acid	NBR	0858	0758	1058	1658	0268	0217	0242	0226	0279	6618	0686-0226	0220	6618	PEEK	
boron trichloride	FEPM	NR	6618	1043	1606	0268	0212	0242	0214	0214	6618	0686-0226	0214	6618	PEEK	316 SS
bromine	FEPM	NR	6618	1043	1606	0268	0212	0242	0214	0214	6618	0686-0226	0220	6618		
brown stock	FEPM	NR	6665	1043	1606	0268	0217	0242	0226	0226	6665	0686-0226	0248	6665		
bunker fuel oil	NBR	0838	0805	1056	1601	0268	0217	0242	0226	0279	0805	0686-0226	0217	0805	PEEK	316 SS
butadiene	FKM	NR	6625	1073	1608	0250	0217	0242	0226	0210	6625	0686-0226	0226	6625	PEEK	BRONZE
butane	NBR	0838	0805	1056	1601	0268	0217	0242	0210	0210	0805	0686-0226	0217	0805	PPS	BRONZE
buttermilk	NBR	NR	6681	1077	1609	0250	0245	0242	0245	0245	6681	0686-0226	0245	6681	PEEK	316 SS

NR: Not Recommended

Consult engineering on blank spaces.

NOTE: Under JFD Design 0838 – Temperatures over 180ºF. Pressure over 2500 psi UTEX recommends 0858

CHEMICAL MEDIA DESCRIPTION	POLYMER TYPE	Recip. MOLDED (LUBRICATED) IFD	SF	SSF	ASF	BRAIDED ADJUSTABLE	BRAIDED NON ADJUSTABLE LUBE	NO LUBE	Centrifugal/Rotary BRAIDED PREMIUM	STANDARD	MOLDED	Valve BRAIDED PREMIUM	STANDARD	Plastic Components	Metal Components
butyl acetate	PTFE	NR	4511	4516	1650	0250	0217	0212	0228	0279	6677,4510	0686-0226	0232	PTFE	BRONZE
butyl alcohol	NBR	0838	0805	1056	1601	0268	0212	0242	0279	0231	0805	0686-0226	0220	PEEK	BRONZE
butylamine	PTFE	NR	4511	4516	1650	0250	0217	0212	0226	0279	6677,4510	0686-0226	0217	PTFE	CAST IRON
butylene	FKM	NR	6625	1073	1608	0268	0217	0242	0226	0210	6625	0686-0226	0217	PEEK	BRONZE
butylene glycol	NBR	0838	0805	1056	1601	0268	0217	0242	0226	0279	0805	0686-0226	0220	PEEK	316 SS
butyl ether	PTFE	NR	4511	4516	1650	0250	0217	0212	0228	0279	6677,4510	0686-0226	0232	PTFE	316 SS
butyric acid	PTFE	NR	4511	4516	1650	0250	0217	0212	0226	0231	6677,4510	0686-0226	0232	PTFE	316 SS
calcium acetate	FEPM	NR	6618	1043	1606	0268	0212	0242	0214	0279	6618	0686-0226	0220	PEEK	BRONZE
calcium carbonate	NBR	0838	0805	1056	1601	0268	0212	0242	0279	0208	0805	0686-0226	0220	PEEK	BRONZE
calcium chlorate	NBR	0858	0758	1058	1658	0268	0217	0242	0228	0279	0805	0686-0226	0220	PEEK	316 SS
calcium chloride	NBR	0858	0758	1058	1658	0255	0217	0242	0279	0208	0805	0686-0226	0220	PEEK	BRONZE
calcium cyanide	NBR	0858	0758	1058	1658	0268	0212	0242	0226	0210	0805	0686-0226	0220	PEEK	BRONZE
calcium hydrogen sulfite	NBR	0858	0758	1058	1658	0268	0217	0242	0228	0279	0805	0686-0226	0220	PPS	316 SS
calcium hydrosulfide	NBR	0858	0758	1058	1658	0268	0217	0242	0228	0279	0805	0686-0226	0220	PEEK	BRONZE
calcium hydroxide, aqueous	NBR	0858	0758	1058	1658	0255	0212	0242	0210	0208	0805	0686-0226	0220	PEEK	316 SS
calcium hypochlorite	EPDM	NR	6665	1081	1610	0255	0212	0242	0226	0214	6665	0686-0226	0220	PEEK	BRONZE
calcium liquors	NBR	0858	0758	1058	1658	0268	0212	0242	0279	0279	0805	0686-0226	0248	PEEK	BRONZE
calcium magnesium chloride	FEPM	NR	6618	1043	1606	0268	0212	0242	0214	0279	6618	0686-0226	0220	PEEK	316 SS
calcium nitrate	NBR	0858	0758	1058	1658	0255	0212	0242	0226	0214	0805	0686-0226	0220	PEEK	BRONZE
calcium phosphate; dibasic, monobasic, tribasic	NBR	0858	0758	1058	1658	0268	0217	0242	0228	0279	0805	0686-0226	0217	PEEK	BRONZE
calcium sulfate, aqueous	NBR	0858	0758	1058	1658	0255	0217	0242	0214	0279	6618	0686-0226	0220	PEEK	BRONZE
carbamate	FEPM	NR	6618	1043	1606	0268	0212	0242	0214	0279	6618	0686-0226	0220	PEEK	316 SS
carbon dioxide, dry	NBR	0858	0758	1058	1658	0268	0217	0242	0226	0210	0805	0686-0226	0217	PEEK	BRONZE
carbon dioxide, wet	NBR	0858	0758	1058	1658	0268	0217	0242	0228	0210	0805	0686-0226	0217	PEEK	BRONZE
carbon disulfide	FEPM	NR	6618	1043	1601	0268	0212	0242	0210	0210	6618	0686-0226	0220	PPS	BRONZE
carbon monoxide	NBR	0838	0805	1056	1608	0268	0212	0242	0226	0210	0805	0686-0226	0217	PEEK	BRONZE
carbon tetrachloride	FKM	NR	6625	1073	1608	0268	0217	0242	0210	0231	6625	0686-0226	0226	PEEK	316 SS
carbonic acid	FKM	NR	6625	1073	1608	0255	0212	0242	0226	0208	6625	0686-0226	0220	PEEK	BRONZE
castor oil	NBR	0858	0758	1058	1658	0268	0249	0242	0279	0231	0805	0686-0226	0217	PEEK	BRONZE
cat cracker slurry	FKM	NR	6625	1073	1609	0250	0245	0242	0214	0245	6625	0686-0226	0245	PEEK	316 SS
catsup	NBR	NR	6681	1077	1604	0268	0217	0242	0245	0217	6681	0686-0226	0220	PEEK	BRONZE
cement	NBR	0842	0809	1068	1633	0268	0217	0242	0214	0217	0809	0686-0226	0220	PEEK	BRONZE
chloric acid	CR	NR	6686	1066	1066	0255	0232	0242	0226	0231	6686	0686-0226	0238	PEEK	NI-RESIST
chlorinated solvents, dry	FKM	NR	6625	1073	1608	0250	0217	0242	0214	0217	6625	0686-0226	0217	PEEK	BRONZE
chlorinated solvents, wet	FKM	NR	6625	1073	1608	0250	0217	0242	0214	0217	6625	0686-0226	0217	PEEK	BRONZE
chlorine dioxide	FKM	NR	6625	1073	1608	0250	0217	0242	0226	0210	6625	0686-0226	0220	PEEK	BRONZE
chlorine, dry	FKM	NR	6625	1073	1608	0250	0217	0242	0210	0231	6625	0686-0226	0217	PEEK	BRONZE
chlorine, wet	FEPM	NR	6618	1043	1606	0250	0217	0242	0226	0210	6618	0686-0226	0217	PEEK	NI-RESIST

NR: Not Recommended
Consult engineering on blank spaces.

NOTE: Under JFD Design 0838 – Temperatures over 180oF, Pressure over 2500 psi UTEX recommends 0858

CHEMICAL MEDIA DESCRIPTION	POLYMER TYPE	Reciprocating / Plunger Pump MOLDED (LUBRICATED) JFD	SF	SSF	ASF	BRAIDED ADJUSTABLE	BRAIDED NON-ADJUSTABLE LUBE	NO LUBE	Centrifugal / Rotary BRAIDED PREMIUM	STANDARD	Valve MOLDED	BRAIDED PREMIUM	STANDARD	Plastic Components	Metal Components
chloroacetic acid	PTFE	NR	4511	4516	1650	0250	0232	0212	0226	0231	6677.4510	0686-0226	0232	PTFE	NI-RESIST
chloroacetone	EPDM	NR	6665	1081	1610	0268	0217	0242	0226	0231	6665	0686-0226	0232	PEEK	BRONZE
chlorobenzene	FKM	NR	6625	1073	1608	0268	0232	0242	0226	0231	6625	0686-0226	0232	PEEK	BRONZE
chloroform	FKM	NR	6625	1073	1608	0268	0212	0242	0214	0279	6625	0686-0226	0220	PEEK	BRONZE
chloropicrin	PTFE	NR	4511	4516	1650	0250	0217	0212	0228	0279	6677.4510	0686-0226	0232	PTFE	NI-RESIST
chlorosulfonic acid	PTFE	NR	4511	4516	1650	0250	0217	0212	0226	0231	6677.4510	0686-0226	0232	PTFE	NI-RESIST
chocolate	NBR	NR	6681	1077	1609	0250	0245	0242	0245	0245	6681	0686-0226	0245	PEEK	316 SS
chromic acid	FEPM	NR	6618	1043	1606	0268	0217	0242	0226	0231	6618	0686-0226	0232	PEEK	316 SS
chromic oxide	FEPM	NR	6618	1043	1606	0268	0217	0242	0226	0231	6618	0686-0226	0232	PEEK	BRONZE
chromium potassium sulfate	NBR	0858	0758	1058	1658	0268	0217	0242	0228	0279	0805	0686-0226	0220	PEEK	BRONZE
citric acid	NBR	0858	0758	1058	1658	0255	0238	0242	0210	0208	0805	0686-0226	0238	PEEK	316 SS
clay slurry	NBR	0842	0809	1068	1604	0268	0217	0242	0226	0214	0809	0686-0226	0217	PEEK	BRONZE
coal-tar	FEPM	NR	6618	1043	1606	0268	0212	0242	0214	0279	6618	0686-0226	0220	PEEK	BRONZE
coal-tar distillate	FKM	NR	6625	1073	1608	0268	0249	0242	0214	0279	6625	0686-0226	0220	PEEK	BRONZE
coconut oil	NBR	0838	0805	1056	1601	0268	0249	0242	0279	0231	0805	0686-0226	0220	PEEK	BRONZE
cod-liver oil	NBR	0838	0805	1056	1601	0268	0249	0242	0279	0231	0805	0686-0226	0220	PEEK	BRONZE
condensate acid < 250 °F	FEPM	NR	6618	1043	1606	0250	0217	0242	0226	0210	6618	0686-0226	0217	PEEK	BRONZE
condensate evaporator < 250 °F	FEPM	NR	6618	1043	1608	0250	0217	0242	0229	0226	6618	0686-0226	0217	PEEK	BRONZE
condensate water < 250 °F	FEPM	NR	6618	1043	1606	0250	0217	0242	0229	0229	6618	0686-0226	0217	PEEK	BRONZE
copper acetate	EPDM	NR	6665	1081	1610	0268	0212	0242	0214	0279	6665	0686-0226	0220	PEEK	316 SS
copper ammonium acetate	EPDM	NR	6665	1081	1610	0268	0212	0242	0214	0279	6665	0686-0226	0220	PEEK	316 SS
copper chloride	NBR	0858	0758	1058	1658	0255	0217	0242	0228	0279	0805	0686-0226	0220	PEEK	NI-RESIST
copper cyanide	NBR	0858	0758	1058	1658	0255	0217	0242	0228	0279	0805	0686-0226	0220	PEEK	316 SS
copper nitrate	NBR	0858	0758	1058	1658	0268	0217	0242	0228	0279	0805	0686-0226	0220	PEEK	316 SS
copper sulfate, < 50%	NBR	0858	0758	1058	1658	0268	0217	0242	0214	0217	0805	0686-0226	0226	PEEK	316 SS
corn oil	NBR	0838	0805	1056	1601	0268	0212	0242	0279	0231	0805	0686-0226	0220	PEEK	BRONZE
cottonseed oil	NBR	0838	0805	1056	1601	0268	0249	0242	0279	0231	0805	0686-0226	0220	PEEK	BRONZE
creosote, coal-tar	NBR	0842	0809	1068	1604	0268	0217	0242	0226	0279	0809	0686-0226	0226	PEEK	BRONZE
cresol	FEPM	NR	6618	1043	1606	0255	0212	0242	0214	0214	6618	0686-0226	0689	PPS	316 SS
crude oil, sour	NBR	0838	0805	1056	1601	0268	0217	0242	0279	0208	0805	0686-0226	0220	PEEK	316 SS
crude oil, sour	FEPM	NR	6618	1043	1606	0268	0217	0242	0226	0210	6618	0686-0226	0217	PEEK	BRONZE
cumene	FKM	NR	6625	1073	1608	0268	0249	0242	0228	0231	6625	0686-0226	0232	PEEK	316 SS
cutting oil	NBR	0838	0805	1056	1601	0268	0217	0242	0279	0231	0805	0686-0226	0220	PEEK	BRONZE
cyanogen	EPDM	0838	0665	1081	1610	0268	0212	0242	0214	0231	6665	0686-0226	0220	PTFE	316 SS
cyclohexane	NBR	0838	0809	1068	1604	0268	0212	0242	0228	0279	0809	0686-0226	0226	PEEK	BRONZE
decahydronaphthalene	FKM	NR	6625	1073	1608	0268	0249	0242	0228	0231	6625	0686-0226	0232	PEEK	316 SS
detergent, H2O solution	NBR	0838	0805	1056	1601	0268	0217	0242	0226	0226	0805	0686-0226	0217	PEEK	BRONZE
diacetone alcohol	EPDM	NR	6665	1081	1610	0268	0212	0242	0279	0231	6665	0686-0226	0220	PEEK	BRONZE

NR: Not Recommended

Consult engineering on blank spaces.

NOTE: Under JFD Design 0838 – Temperatures over 180oF. Pressure over 2500 psi UTEX recommends 0858

CHEMICAL MEDIA DESCRIPTION	POLYMER TYPE	Recip. MOLDED (LUBRICATED) JFD	SF	SSF	ASF	BRAIDED ADJUSTABLE	BRAIDED NON-ADJ. LUBE	BRAIDED NON-ADJ. NO LUBE	Centrifugal BRAIDED PREMIUM	Centrifugal BRAIDED STANDARD	Centrifugal MOLDED	Valve MOLDED	Valve BRAIDED PREMIUM	Valve BRAIDED STANDARD	Plastic Components	Metal Components
diallyl phthalate	PTFE	NR	4511	4516	1650	0250	0217	0212	0228	0279	6677,4510	6677,4510	0686-0226	0232	PTFE	316 SS
dibromoethyl benzene	FKM	NR	6625	1073	1608	0268	0232	0242	0226	0231	6625	6625	0686-0226	0217	PPS	
dibutylamine	FEPM	NR	6618	1043	1606	0268	0217	0242	0226	0279	6618	6618	0686-0226	0217	PPS	CAST IRON
dibutyl Cellosolve adipate	EPDM	NR	6665	1081	1610	0268	0212	0242	0214	0279	6665	6665	0686-0226	0220	PTFE	BRONZE
dibutyl phthalate	PTFE	NR	4511	4516	1650	0250	0217	0242	0228	0279	6677,4510	6677,4510	0686-0226	0232	PTFE	316 SS
dichlorobenzene	FKM	NR	6625	1073	1608	0268	0232	0242	0226	0231	6625	6625	0686-0226	0217	PEEK	BRONZE
diesel oil	NBR	0838	0809	1068	1604	0268	0217	0242	0279	0208	0809	0809	0686-0226	0217	PEEK	BRONZE
diethanolamine	PTFE	NR	6618	1043	1606	0250	0217	0212	0226	0210	6677,4510	6677,4510	0686-0226	0217	PTFE	CAST IRON
diethylamine	PTFE	NR	6618	1043	1606	0250	0217	0212	0226	0210	6677,4510	6677,4510	0686-0226	0217	PTFE	CAST IRON
diethyl carbonate	FKM	NR	6625	1073	1608	0268	0217	0242	0214	0279	6625	6625	0686-0226	0220	PEEK	BRONZE
diethylene glycol	NBR	0838	0809	1068	1604	0268	0217	0242	0279	0208	0809	0809	0686-0226	0238	PEEK	CAST IRON
diethylenetriamine	EPDM	NR	6665	1081	1610	0268	0217	0242	0226	0279	6665	6665	0686-0226	0217	PEEK	CAST IRON
diethyl phthalate	PTFE	NR	4511	4516	1650	0250	0249	0212	0214	0231	6677,4510	6677,4510	0686-0226	0232	PTFE	
diisobutyl ketone	EPDM	NR	6665	1081	1610	0268	0212	0242	0214	0279	6665	6665	0686-0226	0220	PTFE	CAST IRON
dimethylamine	PTFE	NR	6625	1073	1650	0250	0217	0212	0228	0279	6677,4510	6677,4510	0686-0226	0232	PEEK	316 SS
dimethyl formamide	FEPM	NR	6618	1043	1606	0268	0217	0242	0214	0279	6618	6618	0686-0226	0220	PEEK	316 SS
dimethyl phthalate	PTFE	NR	4511	4516	1650	0250	0217	0212	0228	0279	6677,4510	6677,4510	0686-0226	0232	PTFE	
dimethyl terephthalate	PTFE	NR	4511	4516	1650	0250	0217	0212	0228	0279	6677,4510	6677,4510	0686-0226	0232	PTFE	
dinitrochlorobenzene	FKM	NR	6625	1073	1608	0268	0232	0232	0226	0231	6625	6625	0686-0226	0232	PTFE	BRONZE
dioctyl phthalate	PTFE	NR	4511	4516	1650	0250	0217	0212	0228	0279	6677,4510	6677,4510	0686-0226	0232	PTFE	BRONZE
diphenyl	FKM	NR	6625	1073	1608	0268	0217	0242	0214	0279	6625	6625	0686-0226	0232	PPS	BRONZE
Dowtherm® A	FKM	NR	6625	1073	1608	0268	0217	0242	0214	0279	6625	6625	0686-0226	0220	PPS	BRONZE
(Dow Chemical) E	FEPM	NR	6618	1043	1606	0268	0217	0242	0214	0279	6618	6618	0686-0226	0220	PPS	BRONZE
209	FEPM	NR	6618	1043	1606	0268	0217	0242	0214	0279	6618	6618	0686-0226	0220	PPS	BRONZE
epichlorohydrin	PTFE	NR	4511	4516	1650	0250	0249	0212	0226	0231	6677,4510	6677,4510	0686-0226	0232	PTFE	316 SS
ethane	NBR	0858	0758	1058	1658	0268	0212	0242	0226	0210	0805	0805	0686-0226	0217	PEEK	BRONZE
ethanethiol	PTFE	NR	4511	4516	1650	0250	0249	0212	0226	0231	6677,4510	6677,4510	0686-0226	0232	PTFE	CAST IRON
ethanolamine	FEPM	NR	6618	1043	1606	0268	0217	0242	0228	0210	6618	6618	0686-0226	0217	PPS	BRONZE
ethyl acetate	PTFE	NR	4511	4516	1650	0250	0217	0212	0226	0279	6677,4510	6677,4510	0686-0226	0232	PTFE	BRONZE
ethyl alcohol	NBR	0838	0805	1056	1601	0268	0212	0242	0226	0231	0805	0805	0686-0226	0226	PEEK	CAST IRON
ethylamine	EPDM	NR	6665	1081	1610	0268	0212	0242	0214	0279	6665	6665	0686-0226	0232	PTFE	BRONZE
ethylbenzene	FKM	NR	6625	1073	1608	0268	0217	0212	0226	0231	6625	6625	0686-0226	0217	PPS	316 SS
ethylene	NBR	0838	0805	1056	1601	0268	0212	0242	0214	0279	0805	0805	0686-0226	0220	PPS	
ethylene dibromide	FKM	NR	6625	1073	1608	0268	0212	0242	0226	0226	6625	6625	0686-0226	0217	PEEK	BRONZE
ethylene dichloride	FKM	NR	6625	1073	1608	0268	0212	0242	0226	0279	6625	6625	0686-0226	0217	PPS	BRONZE
ethylene glycol	NBR	0838	0809	1068	1604	0268	0217	0242	0279	0208	0809	0809	0686-0226	0238	PEEK	316 SS
ethylene oxide	PTFE	NR	4511	4516	1650	0250	0217	0212	0228	0279	6677,4510	6677,4510	0686-0226	0232	PTFE	316 SS
ethyl ether	PTFE	NR	4511	4516	1650	0250	0217	0212	0228	0279	6677,4510	6677,4510	0686-0226	0232	PTFE	BRONZE

NR: Not Recommended
Consult engineering on blank spaces.
NOTE: Under JFD Design 0838 – Temperatures over 180oF; Pressure over 2500 psi UTEX recommends 0858

CHEMICAL MEDIA DESCRIPTION	POLYMER TYPE	Reciprocating / Plunger Pump							Centrifugal / Rotary			Valve		Plastic Components	Metal Components
		MOLDED (LUBRICATED)				BRAIDED ADJUSTABLE	BRAIDED NON-ADJUSTABLE		BRAIDED		MOLDED	BRAIDED			
		JFD	SF	SSF	ASF		LUBE	NO LUBE	PREMIUM	STANDARD		PREMIUM	STANDARD		
ethyl formate	FEPM	NR	6618	1043	1606	0268	0212	0242	0214	0279	6618	0686-0226	0220	PEEK	BRONZE
fatty acids	FKM	NR	6625	1073	1608	0255	0217	0242	0279	0208	6625	0686-0226	0217	PEEK	316 SS
ferric sulfate, aqueous	NBR	0858	0758	1058	1658	0268	0217	0242	0214	0217	0805	0686-0226	0217	PEEK	BRONZE
ferrous sulfate, aqueous	NBR	0858	0758	1058	1658	0268	0217	0242	0214	0217	0805	0686-0226	0217	PEEK	BRONZE
fluorine, gas, dry < 300 °F	PTFE	NR	4510	4516	1651	0250	0212	0212	0226	0210	4510	0686-0226	0226	PTFE	316 SS
fluosilicic acid	NBR	0858	0758	1058	1658	0268	0217	0242	0226	0214	0809	0686-0226	0217	PEEK	BRONZE
formaldehyde	PTFE	NR	4511	4516	1650	0250	0217	0212	0226	0231	6677,4510	0686-0226	0226	PTFE	BRONZE
formic acid < 160 °F	EPDM	NR	6665	1081	1610	0268	0249	0242	0226	0231	6665	0686-0226	0217	PEEK	BRONZE
Freon 11® (Du Pont)	NBR	0838	0805	1056	1601	0268	0249	0242	0228	0279	0805	0686-0226	0232	PTFE	BRONZE
Freon 113® (Du Pont)	NBR	0838	0805	1056	1601	0268	0212	0242	0214	0279	0805	0686-0226	0238	PEEK	BRONZE
Freon 114® (Du Pont)	NBR	0838	0805	1056	1601	0268	0212	0242	0214	0279	0805	0686-0226	0238	PEEK	BRONZE
Freon 12® (Du Pont)	NBR	0838	0805	1056	1601	0268	0212	0242	0214	0279	0805	0686-0226	0238	PPS	BRONZE
Freon 22® (Du Pont)	CR	NR	6686	1066	1633	0268	0249	0242	0228	0279	6686	0686-0226	0232	PTFE	BRONZE
Freon 502® (Du Pont)	CR	NR	6686	1066	1633	0268	0249	0242	0228	0279	6686	0686-0226	0232	PTFE	BRONZE
fruit juices	NBR	NR	6681	1077	1609	0250	0245	0242	0245	0245	6681	0686-0226	0245	PEEK	BRONZE
fuel oil, #1, #2	NBR	0838	0805	1056	1601	0268	0217	0242	0226	0210	0805	0686-0226	0217	PEEK	BRONZE
#5 light, #5 heavy, #6	FKM	NR	6625	1073	1608	0268	0217	0242	0226	0210	6625	0686-0226	0217	PEEK	BRONZE
fumaric acid	NBR	0858	0758	1058	1658	0268	0217	0242	0226	0231	0809	0686-0226	0232	PPS	BRONZE
furfural	FEPM	NR	6618	1043	1606	0268	0217	0242	0214	0279	6618	0686-0226	0238	PEEK	BRONZE
gas oil	NBR	0838	0805	1056	1601	0268	0217	0242	0226	0279	0805	0686-0226	0217	PEEK	BRONZE
gasoline	NBR	0838	0805	1056	1601	0268	0217	0242	0279	0208	0805	0686-0226	0217	PEEK	BRONZE
gelatin	NBR	NR	6681	1077	1609	0250	0245	0242	0245	0245	6681	0686-0226	0245	PEEK	316 SS
glucose	NBR	NR	6681	1077	1609	0250	0245	0242	0245	0245	6681	0686-0226	0245	PEEK	316 SS
glycerol	NBR	0838	0805	1056	1601	0268	0217	0242	0228	0279	0805	0686-0226	0220	PEEK	BRONZE
glycine	PTFE	NR	4511	4516	1650	0250	0217	0242	0226	0231	6677,4510	0686-0226	0232	PEEK	BRONZE
green sulfate liquor	FEPM	NR	6618	1043	1606	0250	0217	0242	0226	0231	6618	0686-0226	0248	PPS	316 SS
helium	NBR	0838	0805	1056	1601	0268	0217	0242	0226	0279	0805	0686-0226	0217	PEEK	BRONZE
heptane	NBR	0838	0805	1056	1601	0268	0212	0242	0214	0279	0805	0686-0226	0220	PEEK	BRONZE
hexyl alcohol	NBR	0838	0805	1056	1601	0268	0212	0242	0226	0231	0805	0686-0226	0220	PEEK	BRONZE
hydraulic oil, petroleum based	NBR	0838	0805	1056	1601	0268	0212	0242	0226	0210	0805	0686-0226	0238	PEEK	BRONZE
hydrazine	EPDM	NR	6665	1081	1610	0268	0212	0242	0214	0279	6665	0686-0226	0220	PEEK	BRONZE
hydrobromic acid	FEPM	NR	6618	1043	1606	0255	0217	0242	0226	0231	6618	0686-0226	0232	PEEK	NI-RESIST
hydrochloric acid, concentrated	PTFE	NR	4511	4516	1650	0250	NR	NR	0226	0210	6677,4510	0686-0226	0217	PPS	NR
diluted	FEPM	NR	6618	1043	1606	0255	NR	NR	0226	0210	6618	0686-0226	0217	PEEK	NI-RESIST
< 20%, well service	NBR	0858	0758	1058	1658	0268	0217	0242	0214	0279	0805	0686-0226	0217	PEEK	BRONZE
20% - 30%, well service	HNBR	0858	0758	1058	1658	0268	0217	0242	0226	0279	0758	0686-0226	0217	PEEK	316 SS
hydrocyanic acid	FEPM	NR	6618	1043	1606	0255	0217	0242	0226	0231	6618	0686-0226	0232	PEEK	BRONZE
hydrofluoric acid, cold, < 65%	FKM	NR	6625	1073	1608	0268	0217	0242	0226	0210	6625	0686-0226	0217	PPS	BRONZE

NR: Not Recommended
Consult engineering on blank spaces.

NOTE: Under JFD Design 0838 – Temperatures over 180oF, Pressure over 2500 psi UTEX recommends 0858

CHEMICAL MEDIA DESCRIPTION	POLYMER TYPE	Reciprocating / Plunger Pump							Centrifugal / Rotary				Valve		Plastic Components	Metal Components
		MOLDED JFD	MOLDED-LUBRICATED SF	SSF	ASF	BRAIDED ADJUSTABLE	BRAIDED NON-ADJUSTABLE LUBE	NO LUBE	PREMIUM	BRAIDED PREMIUM	STANDARD	MOLDED	PREMIUM	BRAIDED STANDARD		
hydrogen cold, >65%	PTFE	NR	4511	4516	1650	0250	NR	NR	0226	0226	0210	4511	0686-0226	0217	PTFE	NR
hot, <65%	PTFE	NR	4511	4516	1650	0250	0217	0212	0229	0229	0226	4511	0686-0226	0217	PTFE	NI-RESIST
hot, >65%	PTFE	NR	4511	4516	1650	0250	NR	NR	0229	0229	0226	4511	0686-0226	0217	PTFE	NR
hydrogen chloride, gas, dry	NBR	0838	0805	1056	1601	0268	0217	0242	0226	0226	0231	0805	0686-0226	0232	PEEK	BRONZE
hydrogen chloride, gas, dry	EPDM	NR	6665	1081	1610	0268	0212	0242	0214	0214	0279	6665	0686-0226	0220	PEEK	316 SS
hydrogen fluoride, anhydrous	PTFE	NR	4511	4516	1650	0250	0217	0212	0226	0226	0279	4510	0686-0226	0232	PTFE	316 SS
hydrogen peroxide	FEPM	NR	6618	1043	1606	0250	0217	0242	0210	0210	0208	6618	0686-0226	0217	PEEK	316 SS
hydrogen sulfide; dry, cold	FEPM	NR	6618	1043	1606	0268	0217	0242	0226	0226	0231	6618	0686-0226	0232	PEEK	BRONZE
dry, hot	FEPM	NR	6618	1043	1606	0255	0217	0242	0226	0226	0231	6618	0686-0226	0232	PPS	316 SS
wet, cold	FEPM	NR	6618	1043	1606	0255	0217	0242	0226	0226	0231	6618	0686-0226	0232	PEEK	316 SS
wet, hot	PTFE	NR	4511	4516	1650	0255	0217	0242	0226	0226	0231	6677,4510	0686-0226	0232	PTFE	316 SS
hypochlorous acid	FEPM	NR	6618	1043	1606	0255	0217	0242	0226	0226	0231	6618	0686-0226	0232	PEEK	NI-RESIST
isobutane	NBR	0838	1043	1056	1601	0268	0217	0242	0279	0279	0208	0805	0686-0226	0217	PEEK	BRONZE
isopropyl acetate	PTFE	NR	4511	4516	1650	0250	0217	0212	0228	0228	0279	6677,4510	0686-0226	0232	PTFE	316 SS
isopropyl alcohol	FKM	NR	6625	1073	1608	0268	0217	0242	0279	0279	0208	6625	0686-0226	0217	PEEK	BRONZE
isopropyl ether	PTFE	NR	4511	4516	1650	0250	0217	0212	0228	0228	0279	6677,4510	0686-0226	0232	PTFE	BRONZE
jet fuel	NBR	0838	0805	1056	1601	0268	0217	0242	0226	0226	0210	0805	0686-0226	0217	PEEK	BRONZE
kerosene	NBR	0838	0805	1056	1601	0268	0217	0242	0279	0279	0208	0805	0686-0226	0238	PEEK	BRONZE
lacquer	PTFE	NR	4511	4516	1650	0250	0249	0242	0228	0228	0279	6677,4510	0686-0226	0232	PTFE	BRONZE
lactic acid cold	NBR	0758	0758	1058	1658	0255	0217	0242	0226	0226	0231	0809	0686-0226	0238	PEEK	316 SS
hot	FEPM	NR	6618	1043	1606	0255	0217	0242	0226	0226	0231	6618	0686-0226	0238	PEEK	316 SS
lard	NBR	0838	0805	1056	1601	0268	0249	0242	0279	0279	0231	0805	0686-0226	0220	PEEK	BRONZE
latex	NBR	0842	0809	1068	1604	0268	0212	0242	0226	0226	0214	0809	0686-0226	0238	PEEK	BRONZE
ligroin	NBR	0842	0809	1068	1604	0268	0217	0242	0226	0226	0214	0809	0686-0226	0217	PEEK	BRONZE
lime slurry	NBR	0838	0805	1056	1601	0268	0217	0242	0214	0214	0279	0805	0686-0226	0220	PEEK	BRONZE
linoleic acid	FEPM	NR	6618	1043	1606	0255	0217	0242	0226	0226	0210	6618	0686-0226	0217	PPS	316 SS
linseed oil	NBR	0838	0805	1056	1601	0268	0249	0242	0279	0279	0231	0805	0686-0226	0220	PEEK	BRONZE
liquefied petroleum gas	NBR	0838	0805	1056	1601	0268	0217	0242	0279	0279	0208	0805	0686-0226	0238	PEEK	BRONZE
lubricating oil, petroleum base	NBR	0838	0805	1056	1601	0268	0217	0242	0279	0279	0208	0805	0686-0226	0220	PEEK	BRONZE
magnesium chloride	NBR	0858	0758	1058	1608	0268	0217	0242	0228	0228	0214	0805	0686-0226	0220	PEEK	BRONZE
magnesium hydroxide	FKM	NR	6625	1073	1658	0255	0217	0242	0226	0226	0279	6625	0686-0226	0229	PEEK	BRONZE
magnesium sulfate	NBR	0858	0758	1058	1658	0268	0217	0242	0279	0279	0214	0805	0686-0226	0220	PEEK	BRONZE
maleic acid	FEPM	NR	6618	1043	1658	0255	0217	0242	0226	0226	0231	6618	0686-0226	0238	PEEK	BRONZE
maleic anhydride	FEPM	NR	6618	1043	1606	0268	0212	0242	0214	0214	0279	6618	0686-0226	0220	PEEK	316 SS
malic acid	NBR	0858	0758	1058	1658	0255	0217	0242	0226	0226	0231	0809	0686-0226	0232	PEEK	316 SS
manganous chloride	NBR	0858	0758	1058	1658	0255	0217	0242	0228	0228	0279	0805	0686-0226	0220	PEEK	BRONZE
melamine resin	PTFE	NR	4511	4516	1650	0250	0249	0212	0226	0226	0210	6677,4510	0686-0226	0229	PTFE	
mercuric chloride	NBR	0858	0758	1058	1658	0255	0217	0242	0228	0228	0279	0805	0686-0226	0220	PEEK	316 SS

NR: Not Recommended
Consult engineering on blank spaces.

NOTE: Under JFD Design 0838 – Temperatures over 180oF, Pressure over 2500 psi UTEX recommends 0858

CHEMICAL MEDIA DESCRIPTION	POLYMER TYPE	Reciprocating / Plunger Pump							Centrifugal / Rotary			Valve		Plastic Components	Metal Components
		JFD	MOLDED (LUBRICATED) SF	SSF	ASF	BRAIDED ADJUSTABLE	BRAIDED NON-ADJUSTABLE LUBE	NO LUBE	BRAIDED PREMIUM	STANDARD	MOLDED	BRAIDED PREMIUM	STANDARD		
mercury	NBR	0858	0758	1058	1658	0268	0217	0242	0228	0279	0805	0686-0226	0220	PEEK	316 SS
mesityl oxide	PTFE	NR	4511	4516	1650	0250	0217	0212	0228	0279	6677,4510	0686-0226	0232	PTFE	
methane	NBR	0838	0805	1056	1601	0268	0217	0242	0279	0208	0805	0686-0226	0217	PEEK	BRONZE
methyl acetate	PTFE	NR	4511	4516	1650	0250	0217	0212	0228	0279	6677,4510	0686-0226	0232	PTFE	316 SS
methyl alcohol	NBR	0838	0805	1056	1601	0268	0212	0242	0279	0231	0805	0686-0226	0220	PEEK	BRONZE
methyl benzoate	FEPM	NR	6618	1043	1606	0268	0212	0242	0214	0279	6618	0686-0226	0220	PEEK	
methyl bromide	FEPM	NR	6618	1043	1606	0268	0212	0242	0214	0279	6618	0686-0226	0226	PEEK	BRONZE
methyl chloride	FKM	NR	6625	1073	1608	0255	0212	0242	0214	0279	6625	0686-0226	0226	PEEK	BRONZE
methylene bromide	FKM	NR	6625	1073	1608	0268	0212	0242	0214	0279	6625	0686-0226	0226	PEEK	
methylene chloride	PTFE	NR	4511	4516	1650	0250	0217	0212	0228	0279	6677,4510	0686-0226	0232	PTFE	BRONZE
methyl ethyl ketone	PTFE	NR	4511	4516	1650	0250	0249	0212	0210	0231	6677,4510	0686-0226	0217	PTFE	BRONZE
methyl formate	PTFE	NR	4511	4516	1650	0250	0217	0212	0228	0279	6677,4510	0686-0226	0232	PTFE	
methyl isobutyl ketone	PTFE	NR	4511	4516	1650	0250	0249	0212	0210	0279	6677,4510	0686-0226	0217	PTFE	BRONZE
methyl methacrylate	PTFE	NR	6677	1064	1605	0250	0217	0242	0228	0231	6677,4510	0686-0226	0217	PTFE	
methyl propionate	NBR	0858	0758	1058	1658	0250	0217	0242	0226	0279	0805	0686-0226	0217	PEEK	BRONZE
methyl tertiary butyl ether	PTFE	NR	6677	1064	1605	0250	0249	0212	0210	0231	6677,4510	0686-0226	0217	PTFE	
MIL F-25558 (RJ-1)	NBR	0838	0805	1056	1601	0268	0217	0242	0226	0279	0805	0686-0226	0217	PEEK	BRONZE
MIL H-5606	NBR	0838	0805	1056	1601	0268	0217	0242	0226	0208	0805	0686-0226	0217	PEEK	BRONZE
MIL L-7808	FEPM	NR	6618	1043	1606	0268	0212	0242	0226	0279	6618	0686-0226	0217	PEEK	BRONZE
milk	NBR	NR	6681	1077	1609	0250	0245	0242	0245	0245	6681	0686-0226	0245	PEEK	BRONZE
mineral oil	NBR	0858	0758	1058	1658	0268	0217	0242	0279	0231	0805	0686-0226	0220	PEEK	BRONZE
molasses	NBR	NR	6681	1077	1609	0250	0217	0242	0245	0245	6681	0686-0226	0220	PEEK	BRONZE
naphtha, crude	FKM	NR	6625	1073	1608	0268	0217	0242	0279	0208	6625	0686-0226	0238	PEEK	BRONZE
	FKM	NR	6625	1073	1608	0268	0217	0242	0279	0208	6625	0686-0226	0238	PEEK	BRONZE
naphthalene	FKM	NR	4511	1073	1650	0250	0212	0212	0214	0210	6625	0686-0226	0220	PEEK	BRONZE
natural gas,	NBR	0838	0805	1056	1601	0268	0217	0242	0279	0279	0805	0686-0226	0217	PEEK	BRONZE
natural gas, sour	FEPM	NR	6618	1043	1606	0268	0217	0212	0226	0208	0805	0686-0226	0217	PEEK	BRONZE
nickel chloride	NBR	0858	0758	1058	1658	0268	0217	0242	0214	0210	6618	0686-0226	0220	PEEK	BRONZE
nickel sulfate	NBR	0858	0758	1058	1658	0268	0217	0242	0228	0279	0805	0686-0226	0220	PEEK	BRONZE
nitric acid, diluted	FKM	NR	6625	1073	1608	0255	0217	0242	0210	0208	6625	0686-0226	0217	PEEK	316 SS
nitric acid, concentrated	FKM	NR	6625	1073	1608	0255	0217	0242	0210	0210	6625	0686-0226	0217	PPS	316 SS
nitric acid, red fuming	PTFE	NR	4511	4516	1650	0250	0249	0212	0226	0210	6677,4510	0686-0226	0232	PTFE	316 SS
nitrobenzene	PTFE	NR	4511	4516	1650	0250	0232	0212	0226	0231	6677,4510	0686-0226	0232	PTFE	BRONZE
nitrogen gas	NBR	0838	0805	1056	1601	0268	0217	0242	0279	0208	0805	0686-0226	0217	PEEK	BRONZE
nitromethane	PTFE	NR	4511	4516	1650	0250	0217	0212	0226	0217	6677,4510	0686-0226	0217	PTFE	BRONZE
oleic acid	FEPM	NR	6618	1043	1606	0268	0217	0242	0226	0279	6618	0686-0226	0238	PEEK	BRONZE
olive oil	NBR	NR	6681	1077	1609	0250	0245	0242	0245	0245	6681	0686-0226	0245	PEEK	BRONZE
oxalic acid	FKM	NR	6625	1073	1608	0255	0217	0242	0210	0208	6625	0686-0226	0238	PPS	BRONZE

NR: Not Recommended
Consult engineering on blank spaces.
NOTE: Under JFD Design 0838 – Temperatures over 1800F. Pressure over 2500 psi UTEX recommends 0858

CHEMICAL MEDIA DESCRIPTION	POLYMER TYPE	Recip/Plunger MOLDED (LUBRICATED) JFD	SF	SSF	ASF	BRAIDED ADJUSTABLE	BRAIDED NON-ADJ. LUBE	NO LUBE	Centrifugal/Rotary MOLDED	BRAIDED PREMIUM	BRAIDED STANDARD	Valve BRAIDED PREMIUM	Valve BRAIDED STANDARD	Plastic Components	Metal Components
oxygen, gas -10 °F to 200 °F	FEPM	NR	6618	1043	1606	0268	0217	0242	6618	0226	0279	0686-0226	0217	PEEK	BRONZE
oxygen, gas 200 °F to 400 °F	FKM	NR	6625	1073	1608	0268	0217	0242	6625	0226	0279	0686-0226	0217	PEEK	316 SS
ozone	EPDM	NR	6665	1081	1610	0268	0217	0242	6665	0226	0279	0686-0226	0217	PEEK	BRONZE
paint, oil based	NBR	0838	0805	1056	1601	0268	0217	0242	0805	0214	0279	0686-0226	0217	PEEK	BRONZE
palm oil	NBR	0838	0805	1056	1601	0268	0249	0242	0805	0279	0231	0686-0226	0220	PEEK	BRONZE
palmitic acid	NBR	0858	0758	1058	1658	0268	0217	0242	0809	0210	0231	0686-0226	0226	PEEK	BRONZE
paper stock	FEPM	NR	6618	1043	1606	0268	0212	0242	6618	0226	0279	0686-0226	0226	PEEK	BRONZE
paraffin wax, molten	FEPM	NR	6618	1043	1606	0268	0249	0242	6618	0226	0210	0686-0226	0226	PEEK	BRONZE
peanut oil, vegetable	NBR	0838	0805	1056	1601	0268	0249	0242	0805	0279	0231	0686-0226	0220	PEEK	BRONZE
pectin, liquor	NBR	0838	0805	1056	1601	0268	0217	0242	0805	0228	0231	0686-0226	0220	PEEK	BRONZE
pentane	NBR	0838	0805	1056	1601	0268	0212	0242	0805	0214	0279	0686-0226	0220		
perchloric acid	FEPM	NR	6618	1043	1606	0255	0217	0242	6618	0210	0279	0686-0226	0237	PEEK	BRONZE
perchloroethylene	FKM	NR	6625	1073	1608	0268	0238	0242	6625	0214	0231	0686-0226	0220	PEEK	BRONZE
phenol, 10%	FKM	NR	6625	1073	1608	0268	0217	0242	6625	0214	0217	0686-0226	0217	PEEK	316 SS
	FKM	NR	6625	1073	1608	0268	0217	0242	6625	0214	0210	0686-0226	0217	PEEK	316 SS
phenylacetic acid	NBR	NR	6665	1081	1610	0268	0217	0242	6665	0226	0231	0686-0226	0232	PEEK	316 SS
phosphoric acid, concentrated	FKM	NR	6625	1073	1608	0268	0249	0242	6625	0229	0210	0686-0226	0217	PEEK	316 SS
phosphoric acid, diluted	FKM	NR	6625	1073	1608	0268	0249	0242	6625	0226	0210	0686-0226	0217	PEEK	316 SS
phthalic anhydride	EPDM	NR	6665	1081	1610	0268	0212	0242	6665	0214	0279	0686-0226	0220	PEEK	BRONZE
picoline, alpha	EPDM	NR	6665	1081	1610	0268	0217	0242	6665	0226	0279	0686-0226	0238	PEEK	BRONZE
picric acid, H_2O solution	NBR	0858	0758	1058	1658	0255	0217	0242	0809	0226	0231	0686-0226	0238	PEEK	316 SS
picric acid, molten	FEPM	NR	6618	1043	1606	0255	0217	0242	6618	0226	0231	0686-0226	0232	PEEK	316 SS
pine oil	NBR	0838	0805	1056	1601	0268	0217	0242	0805	0279	0279	0686-0226	0238	PEEK	BRONZE
polyethylene glycol	NBR	0838	0805	1056	1601	0268	0217	0242	0809	0226	0208	0686-0226	0238	PEEK	BRONZE
polypropylene slurry	EPDM	0842	0809	1068	1604	0268	0217	0242	0809	0214	0214	0686-0226	0217	PEEK	BRONZE
polyvinyl acetate emulsion	NBR	NR	6665	1058	1658	0268	0217	0242	6665	0214	0217	0686-0226	0217	PEEK	BRONZE
polyvinyl alcohol	NBR	0838	0805	1056	1601	0268	0217	0242	0805	0210	0231	0686-0226	0217	PEEK	BRONZE
potassium bromide	NBR	0858	0758	1058	1658	0255	0212	0242	0809	0228	0208	0686-0226	0220	PEEK	316 SS
potassium carbonate	NBR	0842	0809	1068	1604	0268	0217	0242	0809	0214	0208	0686-0226	0238	PEEK	BRONZE
potassium chlorate	NBR	0858	0758	1058	1658	0268	0217	0242	0805	0228	0279	0686-0226	0220	PEEK	316 SS
potassium chloride	NBR	0858	0758	1058	1658	0255	0217	0242	0805	0210	0208	0686-0226	0238	PEEK	BRONZE
potassium cyanide	NBR	0858	0758	1058	1658	0255	0217	0242	0805	0214	0208	0686-0226	0238	PEEK	316 SS
potassium dichromate	NBR	0858	0758	1058	1658	0268	0217	0242	0805	0228	0279	0686-0226	0220	PEEK	BRONZE
potassium hydroxide, diluted	CR	NR	6686	1066	1633	0255	0212	0242	6686	0208	0208	0686-0226	0238	PEEK	BRONZE
potassium hydroxide, concentrated	FEPM	NR	6618	1043	1606	0255	0217	0242	6618	0210	0208	0686-0226	0238	PEEK	316 SS
potassium nitrate	NBR	0858	0758	1058	1658	0255	0212	0242	0805	0214	0279	0686-0226	0220	PEEK	316 SS
potassium permanganate	EPDM	NR	6665	1081	1610	0220	0212	0242	6665	0214	0279	0686-0226	0220	PEEK	BRONZE
potassium phosphate; *dibasic, monobasic, tribasic*	NBR	0858	0758	1058	1658	0268	0217	0242	0805	0228	0279	0686-0226	0220	PEEK	BRONZE

NR: Not Recommended

Consult engineering on blank spaces.

NOTE: Under JFD Design 0838 – Temperatures over 180oF. Pressure over 2500 psi UTEX recommends 0858

CHEMICAL MEDIA DESCRIPTION	POLYMER TYPE	Reciprocating / Plunger Pump MOLDED (LUBRICATED) JFD	SF	SSF	ASF	BRAIDED ADJUSTABLE	BRAIDED NON-ADJUSTABLE LUBE	NO LUBE	Centrifugal / Rotary BRAIDED PREMIUM	BRAIDED STANDARD	MOLDED	Valve BRAIDED PREMIUM	STANDARD	Plastic Components	Metal Components
potassium sulfate	NBR	0858	0758	1058	1658	0268	0217	0242	0228	0279	0805	0686-0226	0220	PEEK	BRONZE
propane	NBR	0838	0805	1056	1601	0268	0217	0242	0226	0210	0805	0686-0226	0217	PEEK	BRONZE
propionic acid, <150 °F	EPDM	NR	6665	1081	1610	0268	0217	0242	0226	0210	6665	0686-0226	0217	PEEK	BRONZE
propyl acetate	PTFE	NR	4511	4516	1650	0250	0217	0212	0228	0279	6677,4510	0686-0226	0232	PTFE	316 SS
propyl alcohol	NBR	0838	0805	1056	1601	0268	0217	0242	0210	0231	0805	0686-0226	0217	PEEK	BRONZE
propylene	FKM	NR	6625	1073	1608	0268	0217	0242	0226	0210	6625	0686-0226	0217	PEEK	BRONZE
propylene oxide	PTFE	NR	4511	4516	1650	0250	0217	0212	0228	0279	4510	0686-0226	0232	PTFE	316 SS
pulp stock	NBR	0842	0809	1068	1604	0268	0249	0242	0214	0228	0809	0686-0226	0220	PEEK	BRONZE
pyridine	PTFE	NR	4511	4516	1650	0250	0217	0212	0228	0279	6677,4510	0686-0226	0232	PTFE	BRONZE
quenching oil	NBR	0838	0805	1056	1601	0268	0249	0242	0279	0231	0805	0686-0226	0220	PEEK	BRONZE
rapeseed oil	FEPM	NR	6618	1043	1606	0268	0249	0242	0279	0231	6618	0686-0226	0232	PEEK	BRONZE
red liquor	EPDM	NR	6665	1081	1610	0268	0217	0242	0228	0231	6665	0686-0226	0220	PEEK	BRONZE
refrigerator oil	NBR	0838	0805	1056	1601	0268	0249	0242	0279	0231	0805	0686-0226	0220	PEEK	BRONZE
sewage	NBR	0838	0805	1056	1658	0255	0217	0242	0279	0208	0805	0217	0201	PEEK	BRONZE
silver nitrate	NBR	0858	0758	1058	1658	0255	0217	0242	0228	0279	6656	0686-0226	0220	PEEK	316 SS
soap solutions	NBR	0838	0805	1056	1601	0268	0212	0242	0229	0208	0805	0686-0226	0217	PEEK	BRONZE
soda ash	NBR	0858	0758	1058	1658	0268	0212	0242	0214	0217	0809	0686-0226	0217	PEEK	BRONZE
sodium acetate	EPDM	NR	6665	1081	1610	0268	0212	0242	0226	0279	6665	0686-0226	0220	PEEK	BRONZE
sodium bicarbonate	NBR	0858	0758	1058	1658	0268	0217	0242	0226	0214	0809	0686-0226	0217	PEEK	BRONZE
sodium bisulfate	NBR	0858	0758	1058	1658	0268	0217	0279	0228	0279	0809	0686-0226	0220	PEEK	BRONZE
sodium bisulfite, <200 °F	NBR	0858	0758	1058	1658	0268	0217	0242	0279	0231	0809	0686-0226	0217	PEEK	316 SS
sodium carbonate	NBR	0838	0809	1068	1604	0255	0217	0242	0214	0208	0809	0686-0226	0238	PEEK	BRONZE
sodium carbonate >200 °F	FEPM	NR	6618	1043	1606	0255	0217	0242	0226	0214	6618	0686-0226	0226	PEEK	316 SS
sodium chloride	NBR	0838	0809	1068	1604	0268	0217	0242	0226	0279	0809	0686-0226	0217	PEEK	BRONZE
sodium chloride >200 °F	FEPM	NR	6618	1043	1606	0255	0212	0242	0214	0217	6618	0686-0226	0217	PPS	316 SS
sodium cyanide, aqueous	NBR	0858	0758	1058	1658	0268	0212	0242	0214	0279	0805	0686-0226	0220	PEEK	316 SS
sodium dichromate	NBR	0858	0758	1058	1658	0268	0217	0242	0228	0217	0618	0686-0226	0217	PEEK	316 SS
sodium dithionite	FEPM	NR	6618	1043	1606	0268	0249	0242	0214	0217	6618	0686-0226	0217	PEEK	316 SS
sodium hydroxide, diluted	FEPM	NR	6618	1043	1658	0255	0217	0242	0279	0229	6618	0686-0226	0226	PEEK	BRONZE
sodium hypochlorite, 20%	FEPM	NR	6618	1043	1606	0268	0217	0242	0687	0208	6618	0686-0226	0238	PEEK	NI-RESIST
sodium nitrate	FEPM	NR	6618	1043	1606	0268	0249	0242	0214	0279	6618	0686-0226	0220	PEEK	BRONZE
sodium peroxide	FEPM	NR	6618	1043	1606	0268	0212	0242	0279	0208	6618	0686-0226	0220	PEEK	316 SS
sodium phosphate; dibasic, monobasic, tribasic	NBR	0858	0758	1058	1606	0255	0249	0242	0279	0279	0805	0686-0226	0238	PEEK	BRONZE
sodium silicate	NBR	0858	0758	1058	1658	0255	0212	0242	0228	0208	0805	0686-0226	0220	PEEK	BRONZE
sodium sulfate	NBR	0858	0758	1058	1658	0268	0249	0242	0279	0208	0805	0686-0226	0220	PEEK	316 SS
sodium sulfide	NBR	0858	0758	1058	1658	0268	0217	0242	0210	0279	0805	0686-0226	0238	PEEK	BRONZE
sodium sulfite	NBR	0858	0758	1058	1658	0268	0217	0242	0228	0208	0805	0686-0226	0220	PEEK	316 SS
sodium thiocyanate	NBR	0858	0758	1058	1658	0268	0217	0242	0279	0208	0805	0686-0226	0217	PEEK	BRONZE

NR: Not Recommended

Consult engineering on blank spaces.

NOTE: Under JFD Design 0838 – Temperatures over 180oF. Pressure over 2500 psi UTEX recommends 0858

CHEMICAL MEDIA DESCRIPTION	POLYMER TYPE	Reciprocating / Plunger Pump MOLDED (LUBRICATED) JPD	SF	SSF	ASF	BRAIDED ADJUSTABLE	BRAIDED NON-ADJUSTABLE LUBE	BRAIDED NON-ADJUSTABLE NO LUBE	Centrifugal / Rotary BRAIDED PREMIUM	BRAIDED STANDARD	MOLDED	Valve BRAIDED PREMIUM	BRAIDED STANDARD	Plastic Components	Metal Components
sodium thiosulfate	FEPM	NR	6618	1043	1606	0268	0217	0242	0210	0208	6618	0686-0226	0238	PEEK	316 SS
soybean oil	NBR	0838	0805	1056	1601	0268	0249	0242	0279	0231	0805	0686-0226	0220	PEEK	BRONZE
starch	NBR	0858	0758	1058	1658	0268	0268	0242	0214	0217	0809	0686-0226	0217	PEEK	BRONZE
steam, <350 °F	EPDM	NR	6665	1081	1610	0268	0249	0249	0226	0210	6665	0686-0226	0684	PEEK	BRONZE
350 °F - 500 °F	FEPM	NR	6618	1043	1606	0268	0249	0249	0226	0210	6618	0686-0226	0684	PEEK	BRONZE
501 °F - 900 °F	FEPM	NR	6618	1043	1606	0268	0249	0249	0226	0210	6618	0686-0226	0684	NR	BRONZE
>900 °F	NR	NR	NR	NR	NR	NR	NR	NR	0686-0230	0229	NR	0686-0230	0685	NR	BRONZE
stearic acid	FEPM	NR	6618	1043	1606	0255	0217	0242	0210	0208	6618	0686-0226	0238	PEEK	BRONZE
Stoddard solvent	NBR	0858	0758	1058	1658	0268	0217	0242	0210	0279	0809	0686-0226	0226	PEEK	BRONZE
styrene	PTFE	NR	4511	4516	1650	0250	0217	0212	0226	0231	6677,4510	0686-0226	0217	PTFE	BRONZE
sulfite waste liquor	FKM	NR	6625	1073	1608	0268	0217	0242	0226	0279	6625	0686-0226	0248	PEEK	BRONZE
sulfur, molten	FEPM	NR	6618	1043	1606	0268	0249	0242	0229	0226	6618	0686-0226	0226	PPS	316 SS
in water	CR	NR	6686	1066	1633	0268	0212	0242	0214	0279	6686	0686-0226	0232	PPS	316 SS
sulfur chloride	FEPM	NR	6618	1043	1606	0268	0212	0242	0214	0279	6618	0686-0226	0220	PTFE	316 SS
sulfur dioxide, wet	EPDM	NR	6665	1081	1610	0268	0217	0242	0226	0210	6665	0686-0226	0217	PEEK	316 SS
dry	EPDM	NR	6665	1081	1610	0268	0217	0242	0210	0208	6665	0686-0226	0217	PEEK	BRONZE
sulfuric acid, <50%	FEPM	NR	6618	1043	1606	0268	0217	0242	0210	0210	6618	0686-0226	0226	PEEK	NI-RESIST
50% up to 95%	FEPM	NR	6618	1043	1606	0255	0217	0242	0226	0226	6618	0686-0226	0226	PPS	NI-RESIST
≥ 95%	FEPM	NR	6618	1043	1606	0250	0249	0242	0226	0231	6618	0686-0226	0232	PTFE	316 SS
fuming	FEPM	NR	6618	1043	1606	0250	0249	0242	0226	0231	6618	0686-0226	0232	PTFE	316 SS
sulfurous acid	FEPM	NR	6618	1043	1606	0250	0217	0242	0210	0208	6618	0686-0226	0238	PEEK	316 SS
tall oil	PTFE	NR	4516	4516	1650	0250	0217	0212	0226	0210	6677,4510	0686-0226	0226	PTFE	BRONZE
tallow	NBR	0842	0809	1068	1604	0268	0217	0242	0228	0279	0809	0686-0226	0220	PEEK	BRONZE
tartaric acid, aqueous	NBR	0858	0758	1058	1658	0255	0217	0242	0226	0231	0809	0686-0226	0232	PEEK	316 SS
terephthalic acid	PTFE	NR	4511	4516	1650	0250	0217	0212	0226	0231	6677,4510	0686-0226	0232	PTFE	
tetrachloroethane	FKM	NR	6625	1073	1608	0268	0217	0242	0226	0214	6625	0686-0226	0217	PEEK	316 SS
tetrahydrofuran	PTFE	NR	4511	4516	1650	0250	0217	0212	0226	0279	6677,4510	0686-0226	0232	PTFE	BRONZE
theobroma oil	NBR	NR	6681	1077	1609	0250	0245	0242	0245	0245	6681	0686-0226	0245	PEEK	BRONZE
thiols	EPDM	NR	6665	1081	1610	0268	0212	0242	0214	0214	6665	0686-0226	0232		
titanium dioxide	FKM	NR	6625	1073	1608	0268	0212	0242	0214	0279	6625	0686-0226	0217	PEEK	316 SS
titanium tetrachloride	FKM	NR	6625	1073	1608	0268	0212	0242	0214	0210	6625	0686-0226	0220	PEEK	BRONZE
toluene	FKM	NR	6625	1073	1608	0268	0212	0242	0226	0210	6625	0686-0226	0217		
trichloroethane	FKM	NR	6625	1073	1608	0255	0212	0242	0214	0279	6625	0686-0226	0217	PPS	BRONZE
trichloroethylene	FEPM	NR	6618	1043	1606	0268	0212	0242	0214	0279	6618	0686-0226	0220	PEEK	BRONZE
tricresyl phosphate	FEPM	NR	6618	1043	1606	0268	0212	0242	0226	0279	6618	0686-0226	0217	PEEK	BRONZE
triethanolamine	FEPM	NR	6618	1043	1606	0268	0249	0242	0214	0279	6618	0686-0226	0220	PEEK	CAST IRON
tung oil	NBR	0838	0805	1056	1601	0268	0249	0242	0279	0231	0805	0686-0226	0220		

NR: Not Recommended

Consult engineering on blank spaces.

NOTE: Under JPD Design 0838 – Temperatures over 180oF, Pressure over 2500 psi UTEX recommends 0858

CHEMICAL MEDIA DESCRIPTION	POLYMER TYPE	Reciprocating / Plunger Pump — MOLDED (LUBRICATED) JFD	SF	SSF	ASF	BRAIDED ADJUSTABLE	BRAIDED NON-ADJUSTABLE LUBE	NO LUBE	Centrifugal / Rotary BRAIDED PREMIUM	STANDARD	Valve MOLDED	BRAIDED PREMIUM	STANDARD	Plastic Components	Metal Components
turpentine (oil)	NBR	0838	0809	1068	1604	0268	0217	0242	0279	0208	0809	0686-0226	0238	PEEK	BRONZE
urea	NBR	0858	0758	1058	1658	0268	0217	0242	0228	0279	0805	0686-0226	0220	PEEK	BRONZE
urea-formaldehyde resin	PTFE	NR	4511	4516	1650	0250	0249	0212	0226	0210	6677,4510	0686-0226	0229	PTFE	316 SS
varnish	FKM	NR	6625	1073	1608	0250	0212	0242	0214	0279	6625	0686-0226	0220	PEEK	BRONZE
vinegar	FEPM	NR	6618	1043	1606	0255	0212	0242	0214	0279	6618	0686-0226	0220	PEEK	BRONZE
vinyl acetate	PTFE	NR	4511	4516	1650	0250	0217	0212	0226	0231	6677,4510	0686-0226	0229	PTFE	BRONZE
vinyl chloride	PTFE	NR	4511	4516	1650	0250	0217	0212	0226	0279	6677,4510	0686-0226	0231	PTFE	BRONZE
vinylidine chloride	FKM	NR	6625	1073	1608	0268	0212	0242	0214	0279	6625	0686-0226	0220	PTFE	BRONZE
water, chlorinated	NBR	0838	0805	1056	1601	0255	0212	0242	0229	0208	0805	0686-0226	0217	PEEK	BRONZE
fresh	NBR	0838	0805	1056	1601	0268	0217	0242	0208	0201	0805	0686-0226	0238	PEEK	BRONZE
heavy	EPDM	NR	TMC459	1058	TMC459	0255	0280	0280	0280	0280	TMC459	1305-0280	0280	PEEK	BRONZE
salt or sea	NBR	0858	0758	1058	1658	0268	0217	0242	0279	0220	0809	0686-0226	0201	PEEK	316 SS
whiskey	NBR	NR	6681	1077	1609	0250	0245	0245	0245	0245	6681	0686-0226	0245	PEEK	316 SS
white liquor	NBR	0858	0758	1058	1658	0268	0217	0242	0226	0279	0805	0686-0226	0248	PEEK	BRONZE
wine	NBR	NR	6681	1077	1609	0250	0245	0242	0245	0245	6681	0686-0226	0245	PEEK	316 SS
wood pulp stock	NBR	0842	6681	1068	1604	0268	0249	0242	0214	0228	0809	0686-0226	0217	PEEK	BRONZE
wort	NBR	NR	6681	1077	1609	0255	0245	0242	0245	0245	6681	0686-0226	0245	PEEK	BRONZE
xylene	FKM	NR	6625	1073	1608	0268	0217	0242	0226	0231	6625	0686-0226	0217	PEEK	BRONZE
yeast	NBR	0858	6681	1077	1609	0250	0245	0242	0245	0245	6681	0686-0226	0245	PEEK	BRONZE
zinc chloride	NBR	NR	0758	1058	1658	0268	0217	0242	0226	0231	0809	0686-0226	0217	PEEK	316 SS
zinc nitrate	FKM	NR	6625	1073	1608	0268	0212	0242	0214	0279	6625	0686-0226	0220	PEEK	316 SS
zinc sulfate	NBR	0858	0758	1058	1658	0268	0217	0242	0228	0279	0805	0686-0226	0220	PEEK	BRONZE

NR: Not Recommended
Consult engineering on blank spaces.
NOTE: Under JFD Design 0838 – Temperatures over 180oF, Pressure over 2500 psi UTEX recommends 0858

Appendix D

Equivalent Pressure Drop Reference

Discharge of Water

Through Pipes, Orifices, and Valves

Table of Nomenclature

A	= Area of opening, square feet.
C	= Experimental coefficient.
D	= Inside diameter of pipe, feet.
f	= Coefficient of friction.
g	= 32.2 feet per second per second.
h_1, h_2, h_3	= Individual head losses, feet of water.
H	= Effective head pressure, feet of water.
L	= Length of pipe line, feet.
Q	= Flow of water, gallons per minute.
V	= Velocity of water, feet per second.

Frequently it is necessary to determine the quantity of water which will be discharged through a valve or orifice, or through a certain length of pipe. This problem is one which occurs often in hydraulic work, particularly with reference to the flow of water through long pipes lines from a high level to a lower one.

The quantity of water which is discharged into the atmosphere through a pipe, orifice, or valve is dependent upon several variable factors which include flow pressure, size of pipe, type of valve or orifice, condition of pipe wall surface, length of pipe line, etc. The discharge of water into the atmosphere in terms of head loss follows the general formula:

$$H = C\frac{V^2}{2g}$$
Formula No. 1.

Where the value of "C" is usually determined by experiment.

Discharge of Water Through Pipes

In the case of water flowing through pipes, the quantity discharged is dependent upon three major resistance factors:

A. Entrance loss
B. Loss of head due to friction
C. Loss of head due to velocity

The loss of head at entrance depends upon the condition of the pipe end. A well rounded connection will cause the entrance loss to be decreased appreciably, while a square non-projecting pipe will give a "*C*" value of approximately .5 in Formula No. 1. This latter condition is realized in the majority of installations and therefore, the formula for entrance effect is as follows:

$$h_1 = .5 \frac{V^2}{2g}$$ *Formula No. 2.*

The loss of head due to friction is mainly dependent upon the length and condition of the pipe line and may be expressed as

$$h_2 = \frac{f \times L}{D} \times \frac{V^2}{2g}$$ *Formula No. 3.*

In this equation, the friction factor, "*f*", must be determined from experimental results. For rough approximations, the value of "*f*" may be assumed as .015 for brass tubes, .02 for steel pipe, and .04 for cast iron pipe.

The third factor, the loss of head due to velocity, may be expressed by the following formula:

$$h_3 = \frac{V^2}{2g}$$ *Formula No. 4.*

The total loss of head is the sum of the individual losses and may be expressed as follows:

$$H = h_1 + h_2 + h_3 = (1.5 + \frac{f \times L}{D}) \frac{V^2}{2g}$$

As $Q = VA \times 450$

It follows that $Q = 450A \sqrt{\dfrac{2gH}{1.5 + \dfrac{f \times L}{D}}}$ *Formula No. 5.*

Example: How much water will an 8″ steel pipe line (7.981 inside diameter) 3000 feet long discharge per minute at a point which is 300 feet below the inlet?

Given: $D = \dfrac{7.981}{12} = .665\,\text{feet}$

$A = .3474$ square feet

$H = 300$ feet

$2g = 64.4$ feet per second per second.

$f = .02$ (assumed)

$L = 3000$ feet

Then: $Q = 450 \times .3474 \sqrt{\dfrac{64.4 \times 300}{1.5 + \dfrac{.02 \times 3000}{.665}}}$

$Q = 2268$ gallons per minute

Discharge of Water Through Orifices

The coefficient "C" for orifices is dependent upon the condition of the orifice edges, i.e., an orifice having well rounded edges has a "C" value of approximately 1.23, while one having sharp edges gives a "C" value of approximately 2.69.

Discharge of Water Through Globe Valves

The discharge of water through globe valves under a given flow pressure depends primarily upon the type of valve considered. Crane has conducted tests on various types of globe valves in sizes up to and including $2\frac{1}{2}''$ and has determined that the flow through a globe valve is, roughly, one-half of the flow through a short pipe of the same size.

The chart on page 273 gives the discharge of water through pipe sizes from $\frac{3}{8}''$ to $10''$, in lengths from 1 foot to 200 feet; through orifices in sizes up to $10''$; and through valves in sizes up to $2\frac{1}{2}''$ based upon the formulas or tests described previously. For sizes, lengths, or pressures not given on the chart, the respective formulas or ratios may be used.

Note: In all cases, the pressure used to determine the flow through pipes, orifices, or valves, should be the effective head pressure under flow conditions, not the static pressure in the line before flow commences. This statement applies particularly to lines supplied by pumps, etc.

Application of Chart

Example No. 1

What quantity of water will be discharged through 75 feet of $5''$ standard pipe under a discharge gauge pressure of 50 pounds per square inch?

Enter the upper portion of the chart (page 273) at the intersection of the horizontal line for 75 feet of pipe and the line for the 5″ pipe size. Proceed downward to the intersection with the diagonal for 50 pounds gauge pressure and then continue horizontally to the right where the quantity discharged is given as 2600 gallons per minute on the right hand ordinate.

Example No. 2

A 1¼″ globe valve is discharging water into the atmosphere under a discharge gauge pressure of 50 pounds per square inch. What quantity is discharged in gallons per minute?

Enter the upper portion of the chart (page 273) at the black dot on the line for 1¼″ valve size. Proceed downward to the intersection with the diagonal for 50 pounds gauge pressure and then continue horizontally to the left where the quantity discharged is given as 150 gallons per minute on the left hand ordinate.

Discharge of Water (Cont.)
Through Smooth Pipes, Orifices, and Valves

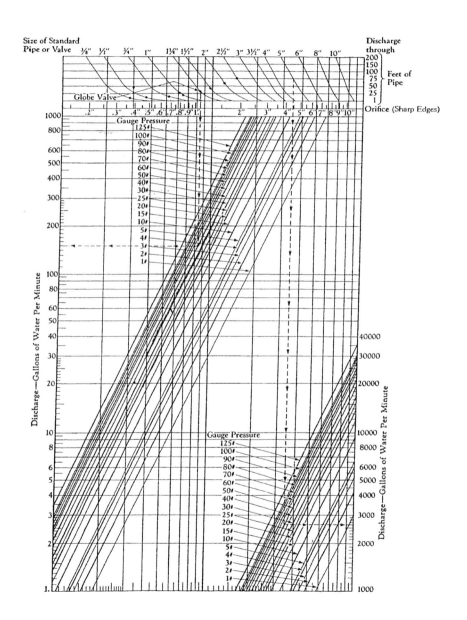

Resistance of Valves and Fittings to Flow of Fluids

When the flow of a fluid in a pipe line is altered by some obstruction such as a valve or fitting, the velocity is changed, turbulence is magnified, and a drop in pressure results. This pressure drop may be insignificant in long lines where it is very small in comparison to the total drop, but when the line is short, the pressure drop through valves and fittings becomes a major item in the total pressure drop value.

The most widely used data on the subject of the resistance of valves and fittings to the flow of water and steam is the information given in Dean Foster's paper on "Effect of Fittings on Flow of Fluids Through Pipe Lines" (published in Vol. 42, 1920, of the *Transactions of the American Society of Mechanical Engineers*). Later tests conducted by others have indicated that the values given by Foster are low for globe valves, angle valves, etc., and that data are needed to establish the correct information to use.

It has been the tendency, and probably is the most practical way, to present the friction values in terms of an equivalent length of the same size of pipe; i.e., the pressure drop caused by a 2-inch elbow is equivalent to approximately the pressure drop caused by five or six feet of 2-inch pipe under the same conditions of flow.

Realizing the need for definite data covering this subject, Crane has conducted pressure drop tests on valves and fittings on both water and steam. These tests were made on 2-inch and 6-inch sizes and conducted under conditions which were thoroughly investigated previous to the tests.

Resistance in Terms of Equivalent Pipe Length

From data given in the tests conducted by Crane and also from information gathered from authentic sources, the chart shown on page 276 has been prepared.

The chart gives the equivalent length of pipe to produce the same pressure drop as a valve or fitting. This additional pipe length should be added to the length of the line in order to determine the total pressure drop.

It has been shown by previous investigators that the drop in pressure through valves, fittings, etc., is some constant multiplied by the velocity head,

$$\frac{V^2}{2g}$$

Therefore: $H_1 = K \dfrac{V^2}{2g}$ where

H_1 = Loss of head in feet
K = Coefficient (values given in table below)
V = Velocity of water, feet per second
$2g$ = 64.4

Representative "K" Values for Various Valves and Fittings

Type	"K"	Authority
Globe valve (fully open)	10.0	Crane tests
Angle valve (fully open)	5.0	Crane tests
Swing check valve (fully open)	2.5	Crane tests
Close return bend	2.2	
Standard tee	1.8	Giesecke & Badgett
Standard elbow	.9 ⎫	⎧ Giesecke & Badgett
Medium sweep elbow	.75 ⎭	⎩ Crane tests
Long sweep elbow	.60	Bulletin No. 2712—University of Texas
45° elbow	.42	Bulletin No. 2712—University of Texas
Gate valve (fully open)	.19 ⎫	
¼ closed	1.15 ⎪	University of Wisconsin tests*
½ closed	5.6 ⎪	and Crane tests
¾ closed	24.0 ⎭	
Borda entrance	.83	"Hydraulics" Daugherty
Sudden enlargement:		
$d_1/d_2 = \frac{1}{4}$	.92	"Hydraulics" Daugherty
$d_1/d_2 = \frac{1}{2}$	.56	"Hydraulics" Daugherty
$d_1/d_2 = \frac{3}{4}$	.19	"Hydraulics" Daugherty
Ordinary entrance	.5	"Hydraulics" Daugherty
Sudden contraction:		
$d_2/d_1 = \frac{1}{4}$	.42	"Hydraulics" Daugherty
$d_2/d_1 = \frac{1}{2}$	.33	"Hydraulics" Daugherty
$d_2/d_1 = \frac{3}{4}$	.19	"Hydraulics" Daugherty

*University of Wisconsin Experimental Station Bulletin Vol. 9. No. 1, 1922

Resistance of Valves and Fittings
to Flow of Fluids

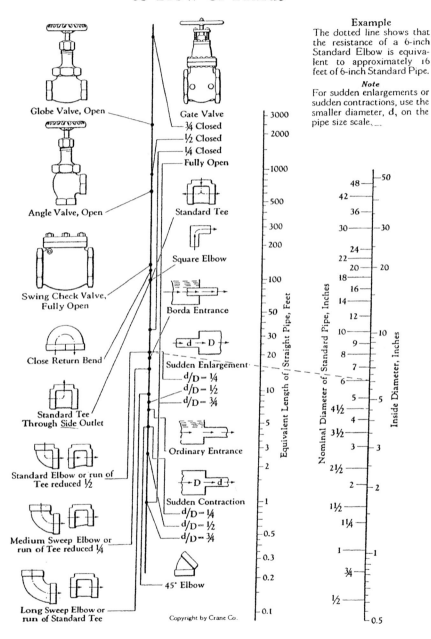

Globe Valve, Open

Angle Valve, Open

Swing Check Valve, Fully Open

Close Return Bend

Standard Tee Through Side Outlet

Standard Elbow or run of Tee reduced ½

Medium Sweep Elbow or run of Tee reduced ¼

Long Sweep Elbow or run of Standard Tee

Gate Valve
¾ Closed
½ Closed
¼ Closed
Fully Open

Standard Tee

Square Elbow

Borda Entrance

Sudden Enlargement
d/D = ¼
d/D = ½
d/D = ¾

Ordinary Entrance

Sudden Contraction
d/D = ¼
d/D = ½
d/D = ¾

45° Elbow

Copyright by Crane Co.

Example
The dotted line shows that the resistance of a 6-inch Standard Elbow is equivalent to approximately 16 feet of 6-inch Standard Pipe.

Note
For sudden enlargements or sudden contractions, use the smaller diameter, d, on the pipe size scale.

3000
2000
1000
500
300
200
100
50
30
20
10
5
3
2
1
0.5
0.3
0.2
0.1

Equivalent Length of Straight Pipe, Feet

Nominal Diameter of Standard Pipe, Inches
48
42
36
30
24
22
20
18
16
14
12
10
9
8
7
6
5
4½
4
3½
3
2½
2
1½
1¼
1
¾
½

Inside Diameter, Inches
50
30
20
10
5
3
2
1
0.5

Pressure Drop of Water Through Schedule 40 Steel Pipe

Pressure Drop of Water per 100 Feet of Schedule 40 Steel Pipe

Columns below are grouped in pairs — **Velocity** (Feet per Second) and **Pressure Drop** (Pounds per Sq. In.) — for successive pipe sizes. Pipe-size labels appear inline in the cell where each size's column begins.

Discharge (Gal/Min)	Veloc	Press	Veloc	Press	Veloc	Press	Veloc	Press	Veloc	Press	Veloc	Press	Veloc	Press	Veloc	Press
.2	1/8″ 1.13	1.52	1/4″ .62	.420												
.3	1.69	4.20	.92	.925	3/8″ .50	.199	1/2″ .32	.065								
.4	2.26	7.70	1.23	1.61	.67	.346	.42	.111								
.5	2.82	11.20	1.54	2.41	.84	.527	.53	.169	3/4″ .30	.045						
.6	3.38	15.75	1.85	3.32	1.01	.747	.63	.238	.36	.064	1″					
.8	4.52	27.40	2.46	5.77	1.34	1.30	.84	.407	.48	.102			1¼″		1½″	
1			3.08	8.80	1.68	1.98	1.06	.605	.60	.152	.37	.049				
2			6.16	33.60	3.37	7.17	2.11	2.18	1.20	.552	.74	.170	.43	.045		
3	2″				5.05	15.30	3.16	4.67	1.80	1.18	1.12	.353	.64	.094	.47	.044
4					6.73	25.70	4.22	7.60	2.40	1.97	1.49	.594	.86	.155	.63	.074
5							5.27	11.15	3.60	2.90	1.86	.902	1.07	.236	.79	.112
6	.57	.046	2½″				6.33	16.10	4.20	4.05	2.24	1.23	1.28	.330	.95	.153
8	.76	.075					8.42	27.60	4.80	6.97	2.98	2.11	1.72	.552	1.26	.263
10	.96	.114	.67	.048	3″				6.01	10.52	3.72	3.08	2.14	.834	1.57	.386
15	1.43	.233	1.00	.099					9.02	22.00	5.60	6.46	3.21	1.76	2.36	.813
20	1.91	.386	1.34	.164	.87	.059	3½″				7.44	11.05	4.29	2.91	3.15	1.35
25	2.39	.581	1.68	.248	1.08	.087	.81	.042	4″				5.36	4.37	3.94	2.02
30	2.87	.804	2.01	.343	1.30	.121	.97	.060					6.43	6.29	4.72	2.91
35	3.35	1.10	2.35	.449	1.52	.158	1.14	.079	.88	.042			7.51	8.25	5.51	3.82
40	3.82	1.37	2.68	.588	1.74	.206	1.30	.100	1.01	.053					6.30	4.78
45	4.30	1.74	3.00	.714	1.95	.251	1.46	.121	1.13	.067					7.08	6.06
50	4.78	2.06	3.35	.882	2.17	.310	1.62	.144	1.26	.080	5″				7.87	7.47
60	5.74	2.96	4.02	1.22	2.60	.429	1.95	.207	1.51	.110						
70	6.69	3.86	4.69	1.53	3.04	.584	2.27	.271	1.76	.150	1.12	.048				
80	7.65	5.03	5.37	2.17	3.48	.762	2.59	.353	2.01	.187	1.28	.063	6″			
90	8.60	6.36	6.04	2.61	3.91	.922	2.92	.446	2.26	.237	1.44	.080				
100	9.56	7.51	6.71	3.23	4.34	1.14	3.24	.527	2.52	.281	1.60	.095	1.11	.039		
125			8.38	4.82	5.42	1.71	4.05	.786	3.15	.438	2.00	.148	1.39	.056		
150			10.06	6.04	6.51	2.35	4.86	1.13	3.78	.602	2.41	.204	1.67	.078		
175			11.73	9.00	7.59	3.20	5.67	1.47	4.41	.820	2.81	.278	1.94	.106	8″	
200					8.68	3.97	6.48	1.92	5.04	1.02	3.21	.346	2.22	.132		
215					9.77	5.02	7.29	2.31	5.67	1.29	3.61	.437	2.50	.166	1.44	.044
250					10.85	6.19	8.10	2.85	6.30	1.59	4.01	.514	2.78	.205	1.60	.055
275					11.94	7.50	8.91	3.44	6.93	1.83	4.41	.622	3.06	.236	1.76	.063
300					13.02	8.47	9.72	4.09	7.56	2.18	4.81	.741	3.33	.280	1.92	.075
325							10.53	4.55	8.18	2.55	5.21	.825	3.61	.329	2.08	.088
350							11.35	5.27	8.82	2.97	5.61	.957	3.88	.362	2.24	.097
375							12.17	6.07	9.45	3.23	6.01	1.10	4.16	.416	2.40	.111
400							12.97	6.89	10.08	3.67	6.41	1.25	4.44	.472	2.56	.127
425	10″						13.78	7.78	10.70	4.15	6.82	1.41	4.72	.534	2.72	.143
450							14.59	8.73	11.33	4.65	7.22	1.50	5.00	.596	2.88	.160
475	1.93	.030							11.96	5.17	7.62	1.67	5.27	.666	3.04	.169
500	2.04	.063							12.59	5.73	8.02	1.85	5.55	.739	3.20	.187
550	2.24	.070	12″						13.84	6.93	8.82	2.74	6.11	.894	3.53	.226
600	2.44	.086							15.10	8.25	9.62	2.67	6.66	1.06	3.85	.270
650	2.65	.101									10.42	3.13	7.21	1.18	4.17	.316
700	2.85	.118	2.01	.048							11.22	3.63	7.77	1.37	4.49	.369
750	3.05	.135	2.15	.055	14″						12.02	4.16	8.32	1.57	4.81	.421
800	3.26	.154	2.29	.062							12.82	4.47	8.88	1.78	5.13	.479
850	3.46	.174	2.44	.070	2.02	.043					13.62	5.05	9.44	2.02	5.45	.511
900	3.66	.194	2.58	.079	2.14	.048					14.42	5.66	10.00	2.26	5.77	.573
950	3.87	.223	2.72	.088	2.25	.053	16″						10.55	2.37	6.09	.638
1,000	4.07	.240	2.87	.098	2.38	.059							11.10	2.63	6.41	.708
1,100	4.48	.274	3.16	.118	2.61	.068							12.22	3.18	7.05	.856
1,200	4.88	.327	3.45	.140	2.85	.081	2.18	.040					13.32	3.78	7.69	1.02
1,300	5.29	.383	3.73	.156	3.09	.095	2.36	.047							8.33	1.13
1,400	5.70	.444	4.02	.180	3.32	.110	2.54	.054	18″						8.97	1.30
1,500	6.10	.511	4.30	.207	3.55	.119	2.73	.062							9.62	1.50
1,600	6.51	.546	4.59	.236	3.80	.135	2.91	.071							10.26	1.70
1,800	7.32	.691	5.16	.298	4.27	.171	3.27	.085	2.58	.048					11.54	2.16
2,000	8.13	.854	5.73	.347	4.74	.211	3.63	.105	2.88	.056	20″				12.83	2.50
2,500	10.16	1.25	7.17	.541	5.92	.309	4.54	.163	3.59	.085			24″			
3,000	12.21	1.80	8.60	.731	7.12	.445	5.45	.221	4.31	.127	3.45	.073				
3,500	14.24	2.42	10.03	.995	8.32	.618	6.35	.300	5.03	.152	4.03	.094				
4,000	16.28	2.99	11.48	1.30	9.49	.792	7.25	.392	5.74	.212	4.61	.122	3.19	.060		
4,500	18.31	3.78	12.90	1.54	10.67	.936	8.17	.497	6.47	.250	5.19	.155	3.59	.060		
5,000	20.35	4.67	14.34	1.89	11.84	1.15	9.08	.572	7.17	.308	5.76	.178	3.99	.073		
6,000	24.42	6.72	17.21	2.73	14.32	1.54	10.88	.824	8.62	.445	6.92	.257	4.80	.100		
7,000	28.50	8.51	20.08	3.72	16.60	1.92	12.69	1.22	10.04	.592	8.06	.350	5.58	.136		
8,000			22.95	4.51	18.98	2.74	14.52	1.36	11.43	.734	9.23	.457	6.38	.178		
9,000			25.80	5.70	21.35	3.47	16.32	1.72	12.92	.910	10.37	.536	7.19	.225		
10,000			28.63	7.04	23.75	4.09	18.16	2.12	14.37	1.13	11.53	.663	7.96	.278		
12,000			34.38	9.36	28.49	6.18	21.80	3.09	17.23	1.65	13.83	.954	9.57	.371		
14,000					33.20	8.40	25.42	4.16	20.10	2.07	16.14	1.20	11.18	.505		
16,000							29.05	5.44	22.96	2.71	18.43	1.57	12.77	.660		

Index